Human Biology

Laboratory Manual

Second Edition

Revised Printing

D0147007

Marty Lowe

Bergen Community College

Contents

Name: Leslie Castillo

I. GENERAL SAFETY PRECAUTIONS AND PROCEDURES

A. Laboratory instructions are given at the beginning of the laboratory class, it is essential that you arrive to laboratory on time.

B. In order to prevent injury to yourself or your laboratory partner(s), please read all experiments before coming to class. If you do not understand a particular portion of the experiment ask your laboratory instructor to clarify the procedure(s). Laboratory class is not a time to improvise.

C. The performance of unauthorized experiments is forbidden.

D. No eating or drinking is allowed in the laboratory.

E. All non-essential items (coats, books, bookbags, etc.) should be stored away from the laboratory table.

F. If you have long hair pull it back to prevent it from becoming a fire hazard.

G. Remove contact lenses when working with chemical reagents.

H. Note the location of the fire extinguisher and eye wash station in your laboratory.

I. Report any injury to your instructor then proceed to the nurse's office.

J. Never remove any equipment or materials from the laboratory.

K. Dispose of all biological and chemical materials in a proper manner. Use the bio-hazard bags when necessary.

L. Use caution when handling caustic agents (strong acids and/or bases).

M. Wash your hands thoroughly before leaving the laboratory.

N. Protective eyeware, lab coats and gloves should be worn by all students performing or observing experiments.

O. Pregnant students and those who suffer from hypertension or any other known or suspected condition, should notify the laboratory instructor prior to performing laboratory exercises that involve cardiovascular stress.

P. When performing a dissection, please adhere to the following guidelines:

1. Work in a well ventilated room. Avoid breathing in the preservative vapors for extended periods. Notify your instructor if you start to feel lightheaded or dizzy.

2. Wear lab coats, gloves, and safety glasses to protect your clothing, skin, and eyes from the chemical preservatives.

3. Wash your specimen with running water prior to dissecting. This will remove any excess preservative and reduce its irritating effects.

4. Extreme care should be observed when using the dissection instruments. Try to use a blunt-edged probe or your fingers as much as possible, this will lessen the likelihood of injuring yourself. If you

cut or puncture yourself, notify your instructor and proceed to the nurse's office.

    5. Clean all dissecting instruments, after use and dry them.

Q. When working with Blood and/or other Body Fluids, please adhere to the following guidelines.

    1. Protective eyeware, laboratory coats, and gloves should be worn during procedures that involve the handling of all specimens or contaminated instruments.

    2. All skin surfaces should be washed immediately if contacted by blood or other body fluids.

    3. Never perform pipetting by mouth. The laboratory is equipped with mechanical pipetting devices.

    4. Only disposable lancets and needles should be used during the laboratory procedure. Dispose of the instruments in specially marked puncture-proof containers.

    5. Dispose of non-reusable equipment in bio-hazard bags located in the laboratory.

    6. Clean all laboratory equipment with a 1:10 dilution of household bleach.

    7. Wash your hands thoroughly, immediately after finishing the laboratory exercise.

R. When performing microbiological experiments, please adhere to the following guidelines.

    1. Notify your instructor if a culture is spilled. Following your instructor's directions for clean-up.

    2. When pipetting always use a pipette pump. Never pipette by mouth.

    3. All disposable materials should be deposited in the bio-hazard bags.

    4. Inoculation loops and needles must be sterilized before returning them to their storage container.

    5. Laboratory tables should be cleaned with a 1:10 bleach solution or commercial cleanser.

    6. Wash your hands thoroughly before leaving the laboratory class.

S. When working with chemical reagents, please adhere to the following guidelines.

    1. Report all accidents to your instructor.

    2. Make sure that the opening of all test tubes never point toward yourself or another person, when heating it.

    3. Flush eyes or skin with water for fifteen minutes if contacted with any reagent. Notify your instructor immediately.

    4. Never inhale fumes directly, rather "waft" the fumes toward you if directed to note an odor.

    5. Seal test tubes and flasks with the appropriate stoppers prior to mixing the contents. Don't use your fingers as stoppers.

    6. Recap bottles immediately after use. Leaving them open may effect the reactivity of the chemical.

    7. Use only one spatula for each chemical. Do not mix them up.

    8. Never pipette by mouth.

    9. Never return any unused portion to its original container.

    10. Never boil anything with a cap or stopper on it.

**Draw a map of your lab and mark the location of the items found
in your lab. An example of a lab diagram is shown on p. 7.**

LABORATORY HUNT

To familiarize yourself with the lab take a few minutes to find the items listed below and mark them on the blank diagram of the laboratory on the next page.

a. Fire Extinguisher
b. Eye Wash Station
c. Broken Glass Disposal Container
d. Biohazard Disposal Container
e. Trash Barrel
f. Beakers, Test Tubes, and Flasks
g. Microscopes
h. Lens Cleaner and Lens Paper
i. Coat Rack
j. Exits
k. Hand Washing Sinks
l. Hot Plates
m. Experiment Set Ups
n. Glass and Hand Soap
o. Test Tube Brushes
p. Dyes for Cells
q. Clean Glass Slides
r. Cover Slips
s. Used Slide Disposal Container
t. Sharps Disposal Container
u. Bunsen Burners and Flint Lighters
v. Dissection Tools
w. Disposable Pipettes
x. Sterile Loops and Swabs
y. Propipettes
z. Water Bath

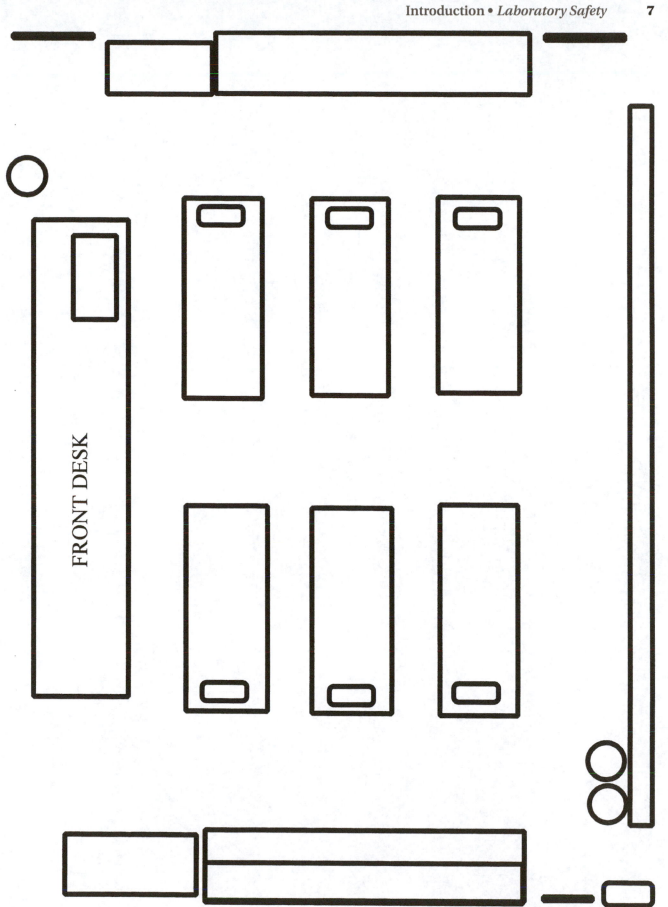

Microorganisms in Human Biology

I. INTRODUCTION

Microorganisms (viruses, chlamydia, rickettsia, bacteria, cyanobacteria, fungi, molds, algae, and protozoa) are found everywhere in nature; from ocean depths to mountain peaks, from swamps to deserts, and from frigid polar regions to the super-heated water escaping from vents on the ocean floor. The environment where an organism is normally found is presumably the environment to which the organism is best adapted, i.e., that environment is the organism's ecological niche.

While individual microorganisms have particular ecological niches, they are not necessarily confined to those niches. In fact, a microorganism transferred to a new niche by accident or deliberate transfer may even find the new environment more adventitious than the environment that is its normal niche. The ability to tolerate, or even to take advantage, of new environments, coupled with the fact that microorganisms are very small and exist virtually everywhere, means that special procedures are necessary for handling them effectively. While more detail and discussion of why organisms exist where they do, this lab exercise can begin to demonstrate just how ubiquitous microorganisms really are.

II. COLONY MORPHOLOGY OF BACTERIA IN OUR ENVIRONMENT

Procedure

1. Groups of four (4) students should obtain five (5) nutrient agar plates from the supply provided and handle them as described below.
    a. Open one plate and expose it to the air for 30 minutes. You may use the lab, the restrooms, or outside.
    b. Gently press some coins to the agar surface.
    c. One individual should kiss or put their tongue on a plate.
    d. Each student should gently press a finger to the plate.
    e. Swab some surface (other than a laboratory desk top) with a sterile swab, then streak (rub) the swab onto the remaining agar plate. Dampen the swab in sterile water to pick up as much bacteria as possible. You may use your cell phone, the doorknobs or items in the restroom. If you go to the cafeteria be sure to get permission from the cafeteria manager.
    Incubate plates (lid down, or inverted) at 37° C until the next lab period.
2. After incubation, observe your plates for different colony types. Focus upon colony surface form and texture, elevation, and margin (see Figure 1.1), and describe several colony types in Table 1.1 on the next page.
3. Practice making other discriminating observations like counting the number of colonies of each type on a given plate, and comparing colony types on the different plates.

From *Laboratory Manual for General Microbiology*, Third Edition by James E. Urban. Copyright © 1996 by James E. Urban. Used with permission of Kendall/Hunt Publishing Company.

AGAR MEDIA

1.   PLATES (Colonies)
    a) *General surface form*

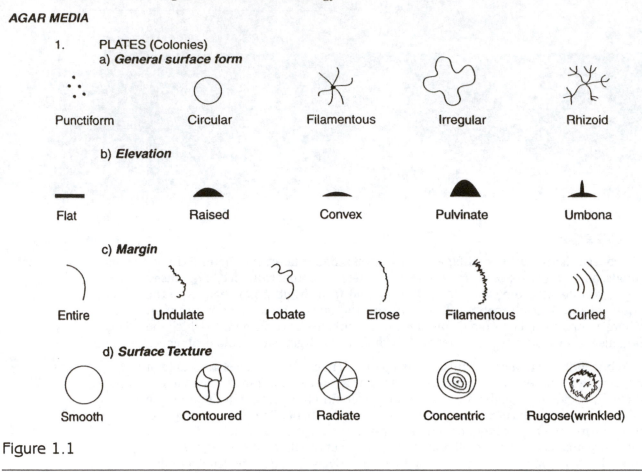

Figure 1.1

Table 1.1 Summary of Microorganism Omnipresence Data			
Environment Sampled	*Colony Characteristics*	*Colony Color*	*Number of Colonies*

From *Laboratory Manual for General Microbiology,* Third Edition by James E. Urban. Copyright ©1996 by James E. Urban. Used with permission of Kendall/Hunt Publishing Company.

III. OUR NORMAL FLORA

Procedure

1. Work with your lab partner. Choose one partner's hand as the test subject. That person must not touch anything with their hand until the experiment is over.
2. Obtain four Agar Plates and four sterile swabs.
3. Label all plates with your initials and the date, around the edge, then:
4. Label the first plate "hand only" at the bottom of the plate around the edge.
5. The second plate should be labeled "warm water rinse"
6. The third plate should be labeled "soap and water"
7. The fourth plate should be labeled "alcohol rinse":
8. Using a sterile swab and sterile water, wet the swab, gently remove excess water on the side of the water test tube, and swab your lab partner's hand. Get the fingers as well as the palm. Gently roll the swab over the agar plate. Do not break the agar. Cover as much of the agar as possible. Dispose of the swab in the orange biohazard bag. Dispose of the sterile water.
9. Have the test subject rinse the same hand under running warm water. Do not let them turn on the water. Get another sterile swab and swab the water off their hand (swab the same places as before). Gently swab the surface of the agar on the second plate.
10. Using the same hand again and the running warm water, squeeze soap onto the test subject's hand and have them rub it in using only that hand. Get a third sterile swab and gently swab the water from their hand and swab the surface of the third agar plate.
11. Finally rinse the test subject hand with alcohol and using a fourth sterile swab remove the liquid from their hand and gently swab the surface of the fourth agar plate.
12. Incubate the plates in an inverted position for 24 hours at 37 degrees Celsius. Next week examine your plates. Draw and describe your results.

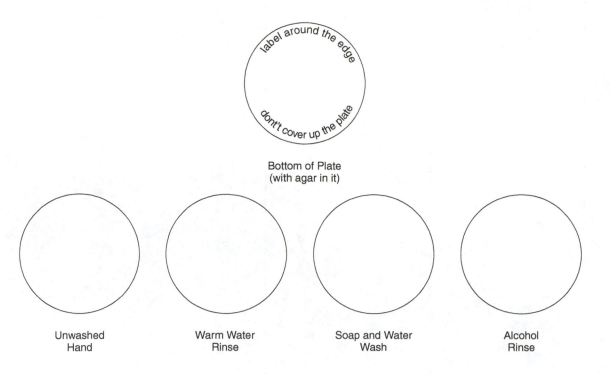

label around the edge

don't cover up the plate

Bottom of Plate
(with agar in it)

Unwashed   Warm Water   Soap and Water   Alcohol
Hand   Rinse   Wash   Rinse

Introduction to Anatomy

I. INTRODUCTION

To the beginning students anatomy is a lot like a foreign language. New vocabulary is used to describe structures and directions with which you are already familiar. The techniques you have used in other courses to study vocabulary will be helpful here. Write the words so you can spell them. Use them in sentences to help remember their meanings. Make flash cards so you and your colleagues can drill each other on the meanings of these words. Most of them are a bit hard to fit into ordinary conversation, (did you realize that the distal portion of your slip is protruding from under your skirt?) but in the next few weeks you will use these terms daily in your lab work, and eventually they will become familiar. Today, however, you will be quite dependent on the glossary at the end of this exercise as you begin to learn anatomical vocabulary.

II. DIRECTIONS, PLANES AND REGIONS

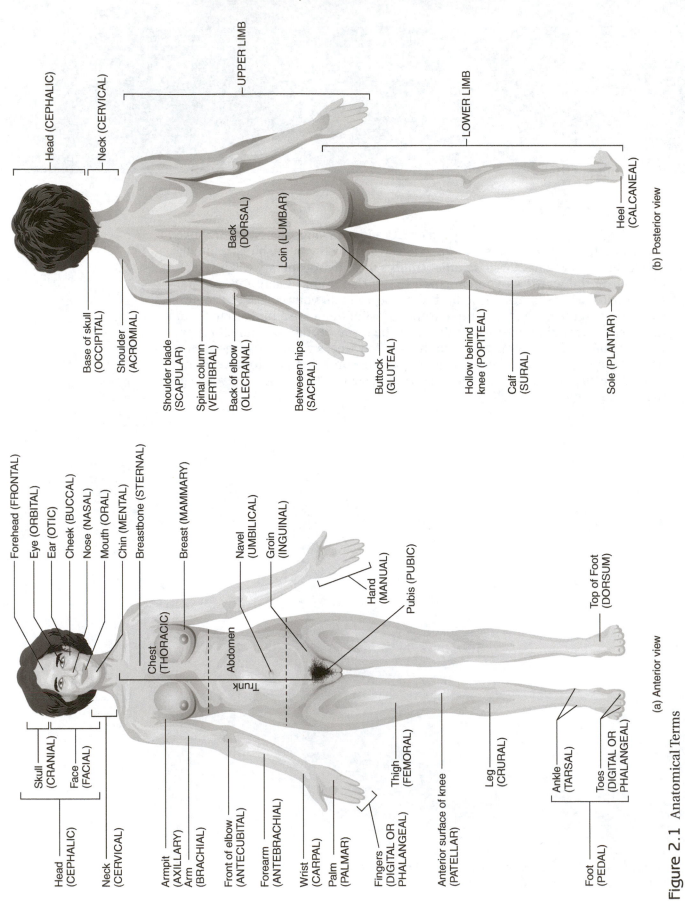

Figure 2.1 Anatomical Terms

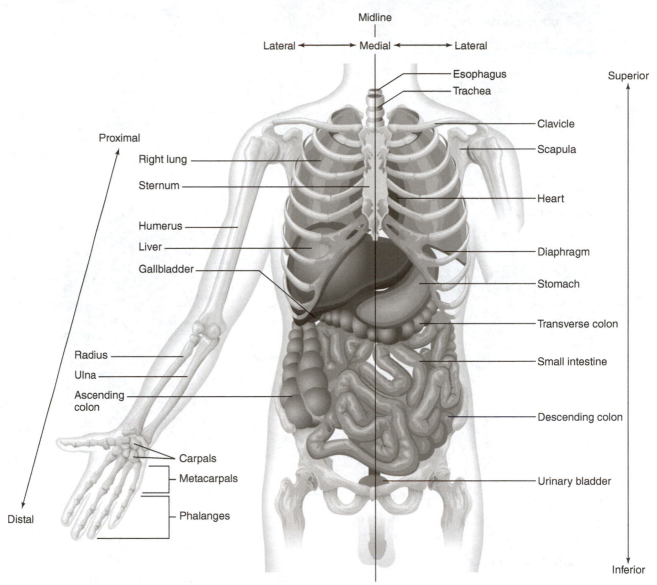

Figure 2.2 Anatomical Terms and Positional Terms

DORSAL BODY CAVITY

VENTRAL BODY CAVITY

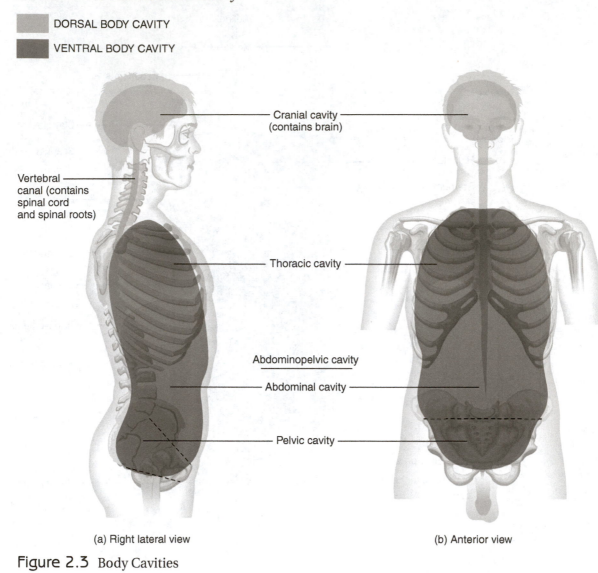

Cranial cavity
(contains brain)

Vertebral canal (contains spinal cord and spinal roots)

Thoracic cavity

Abdominopelvic cavity

Abdominal cavity

Pelvic cavity

(a) Right lateral view

(b) Anterior view

Figure 2.3 Body Cavities

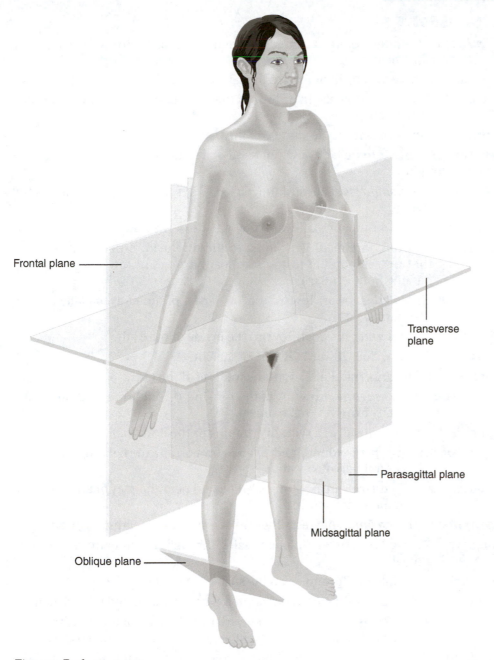

Frontal plane

Transverse plane

Parasagittal plane

Midsagittal plane

Oblique plane

Figure 2.4 Body Planes

III. GLOSSARY

anterior   (L. *ante,* before) Located before or toward the front. In tetrapods, the head end.

caudal   (L. *cauda,* tail) To or toward the tail in a tetrapod.

cranial   (Gr. *kranion,* skull) To or toward the head end of the body. Pertaining to the brain case.

deep   (inside) Away from the surface of the body.

distal   (L. *distare,* to be far away) That end of a structure most distant from its origin.

dorsal   (L. *dorsum,* back) Toward the back, made up of the spinal and cranial cavities.

external   (L. *ex,* from, out of) Toward the outside of the body.

frontal   (L. *frons,* forehead) Pertaining to the forehead. A plane of the body that passes parallel to the forehead; a median longitudinal plane passing from left to right.

inferior   (L. *inferior,* lower) Toward the lower part of the body, usually used in human anatomy.

internal   (L. *internus,* internal) Toward the inside of the body.

lateral   (L. *latus,* side) Toward the side of the body.

medial   (L. *medius,* middle) Toward the midline of the body.

midsagittal   (L. *sagitta,* arrow) A plane passing through the midline of the body from ventral to dorsal.

parietal   (L. *paries,* wall of a room) Pertaining to the body wall.

peripheral   (Gr. *peri,* around, and *pherein,* to carry) External, away from the central nervous system.

posterior   (L. *posterior,* nearer the rear) Located near the rear of an animal. In tetrapods, the tail end.

proximal   (L. *proximus,* nearest) The end of a structure nearest its origin.

sagittal   (L. *sagittus,* arrow) A plane passing through the body from ventral to dorsal.

superficial   (outside) Toward or on the surface of the body.

superior   (L. *superior,* higher) A direction toward the head end of a human.

transverse   (L. *trans,* across, and *vertere,* to turn) A plane of the body crossing its longitudinal axis at a right angle. A cross-sectional plane.

ventral   (L. *venter,* belly) A direction toward the belly surface, made up of the thoracic and abdominopelvic cavities.

visceral   (L. *viscera,* entrails) Pertaining to the inner part of the body rather than the body wall.

Use the following terms to complete the sentences below. Use your text, the glossary, diagrams, class notes, and *common sense* to help you.

superior	inferior	anterior	posterior
pectoral	popliteal	antebrachial	cervical
lateral	medial	internal	external
distal	proximal	peripheral	parietal
plantar	axillary	lumbar	femoral
transverse	visceral	sagittal	midsagittal
frontal	ventral	dorsal	
caudal	occipital	cranial	
superficial	brachium	deep	

1. The ears are _____ to the eyes.

2. The lungs are _____ to the stomach and _____ to the head.

3. The intestines are _____ to the abdominal wall.

4. The breasts are _____ to the lungs.

5. If an incision were made through the heart, cutting it into equal right and left halves, the incision would pass through the _____ plane.

6. The vertebral column is _____ to the kidneys.

7. The knee is _____ to the foot. The elbow is _____ to the shoulder.

8. An incision made so that it divides the brain into upper and lower sections passes through the _____ plane.

9. A cut through the exact midline of the body, made so it is divided into exact right and left halves, passes through the _____ plane.

10. The thoracic wall is lined with _____ membrane.

11. The back of the head is the _____ region.

12. The _____ region is the thigh.

13. The upper part of the arm is the _____ region.

14. The lower back is the _____ region.

15. The underarm is the _____ region.

16. The neck is the _____ region.

17. Behind the knee is the _____ region.

18. The bottom of the foot is the _____ surface.

19. The forearm is the _____.

20. The chest lateral to the sternum is the _____ region.

IV. BODY CAVITIES

List the body cavities where these organs are located:

Organ	*Cavity*
21. brain	_____
22. heart	_____
23. lungs	_____
24. stomach	_____
25. uterus	_____

Complete the following sentences.

26. When a neurosurgeon performs brain surgery, she is operating in the _____ cavity.

27. When cerebrospinal fluid is removed from around the spinal cord it is removed from the _____ and _____ cavities.

28. Would the removal of an appendix be abdominal or thoracic surgery? _____

29. Reproductive and urinary organs are found in the _____ cavity.

30. When a doctor performs heart surgery, he is operating in the _____ and _____ cavities.

31. The _____ is the boundary between the thoracic and abdominopelvic cavities.

V. ANIMALS AND PEOPLE

There is just one small problem with all of this. We will be using a pig or rat as our dissection animal. The animal in which we are really interested is ourselves. As you would expect from members of the same class, Mammalia, the structure of our bodies is really quite similar. We have all the same parts in basically the same places. The big difference is that animals walk on four legs, and we usually walk on two legs. As long as we confine ourselves to a study of one or the other animal we are fine, and there is no confusion. But we will be studying both, and there is plenty of confusion. For example, we walk around with our **ventral** surface facing forward. An animal walks around with its **ventral** surface facing the ground. In humans, **ventral** means the same as **anterior.** But in animals, **ventral** means the same as **inferior,** and **anterior** means toward the head end. Usually the terms **superior** and **inferior** are reserved for humans, and human anatomists usually but not always use **dorsal** and **ventral** instead of **posterior** and **anterior.** Quite often, the terms **cranial** and **caudal** are used to indicate the head and tail end of four-legged animals. These terms are used less often to describe positions in the human body.

Now that you are totally confused, try the following exercise.

32. In the pig the most anterior part of the body is the _____.

33. In humans, the terms cranial and _____ have approximately the same meaning.

34. The part of the pig's body which faces the ground is the _____ side.

35. As you make an incision toward the pig's tail end you are cutting in the _____ or _____ direction.

36. In humans the blood vessel which brings blood from the head to the heart called the superior vena cava. In the pig this vessel is the _____ vena cava.

To really test yourself draw a picture of a dog or cat and label the ventral, dorsal, anterior, and posterior portions. Remember, your posterior refers to your backside and anterior refers to the front or face side of the body. Dorsal is the cavity that holds the brain and spinal cord. The ventral cavity makes space for the internal organs in the trunk.

VI. REVIEW

Label:

Ventral cavity
Dorsal cavity
Thoracic cavity
Abdomen
Pelvic cavity

Abdominopelvic
Pericardial cavity or mediastinum
Cranial cavity
Vertebral cavity

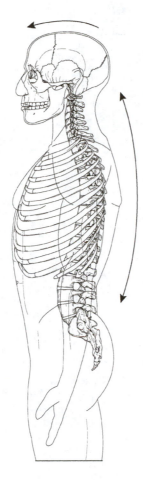

Figure 2.5

Label:

medial, lateral, superior, inferior

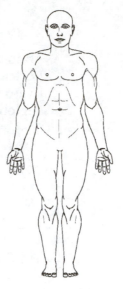

Label:

cranial, spinal or vertebral, dorsal and posterior

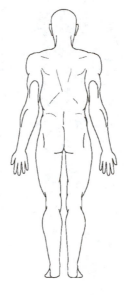

Label:

anterior , posterior, ventral, dorsal, thoracic, abdominopelvic

Figure 2.6

Label: distal and proximal

Which joint is both distal and proximal?

_____

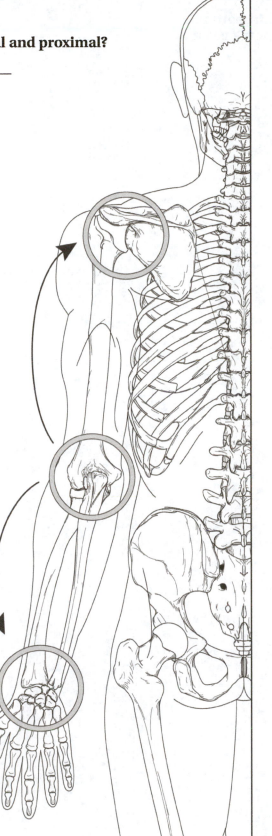

Figure 2.7

Label the planes:

frontal, midsagittal, coronal, or transverse, sagittal

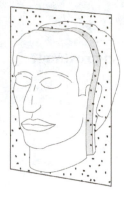

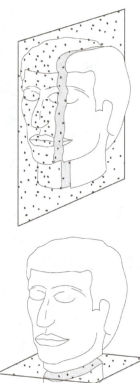

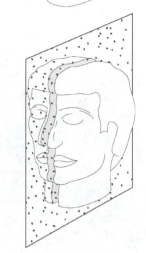

Figure 2.8

Compound Light Microscope

I. INTRODUCTION

The compound light microscope is a valuable tool for any biology student. The following discussion should allow the student to feel comfortable with the use and principles of light microscopy.

II. THE PARTS OF THE COMPOUND LIGHT MICROSCOPE

The **ocular** lenses, or eyepieces, are located at the top of the scope. Notice that the interpupillary distance can be adjusted for each user of the scope. Also notice that the left eyepiece is focusable to adjust for differences in the user's eyes. Focus the specimen through the right eyepiece while the left eyepiece is covered. Next, focus the left eyepiece by covering the right eyepiece and turning the left eyetube until the image is in clear focus. The oculars magnify the image by a factor of 10X; this is the second magnification. The first magnification is performed by the **objective** lenses. There are four objective lenses located on a **revolving nosepiece.** The first is the **scanning** lens which has a magnification of **4X.** The second objective lens is the **low** power objective with a magnification of **10X.** The third lens is the **high-dry** objective lens which has a magnification of **40X.** The fourth lens is the **oil immersion** lens with a magnification of **100X.** The **total magnification** of the image being viewed is determined by multiplying the ocular magnification by the magnification being used. Total magnification notations should be made with all drawings made during lab.

Slides are held on the scope by the use of **slide clips** seen on the surface of the **stage.** Carefully place slides in these holders; do not allow the clip to snap onto the slide as this may cause chipping of the slide. Slides being viewed are moved about the stage by the use of **vertical** and **horizontal knobs.** Do not push slides around the stage; this will strip the gears of the controls.

Beneath the stage is the **condenser.** The purpose of the condenser is to direct the light rays onto the specimen being viewed. The condenser should always be in an up position, which is just beneath the stage. The **iris diaphragm** is part of the condenser; you will see this as a black lever that can be moved in a left to right fashion. The iris diaphragm controls the amount of light entering the condenser. Less light into the condenser means less light on the specimen which will cause the specimen to have more contrast. More light leads to less contrast especially when working with very small or very thin specimens. You may have to adjust the iris diaphragm for each specimen viewed.

Focusing controls are located on the back base of the microscope. The **coarse focus** control is the larger knob located next to the base. Notice the nosepiece moves a considerable amount when using this control. The coarse focus is only used with the scanning or low power lens. The **fine** focus control is the smaller outside knob; this is used to fine-tune the image until it is clear. The microscope is said to be **parfocal** which means the image stays relatively in focus when changing from one magnification to another. For this reason, when changing objective lenses you should only have to adjust the fine focus control.

Resolution is being able to see two separate objects as two distinct objects. In order to see close objects as separate objects, light must pass between them. **Resolving power** is a measure of the distance between these objects. The better the resolving power of the microscope, the less distance between the objects which in turn means the objects can be very small. The equation for resolving power is as follows:

$$RP = \frac{\text{wavelength}}{2 \times NA}$$

NA stands for **numerical aperture;** this number is usually printed on the lens. It is a measure of the cone of light entering the objective lens as seen below:

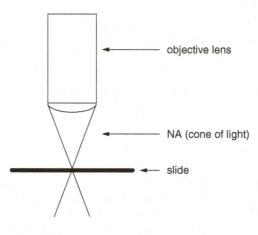

In order to view small objects a small value for RP is desired. To obtain a small number, a small value for wavelength can be divided by a larger number. For example, if 2 is divided by 4, the result is 0.5. If the wavelength used is small and is divided by a larger value for NA, a small number for RP will be obtained. Blue wavelengths of light are smaller than other light rays. Notice that there is a blue filter on top of the light source. Having the condenser closer to the stage will widen the cone of light as far as possible giving a large value for NA. Using these principles we have obtained the best resolution possible for viewing microorganisms.

Notice as higher magnification objectives are used, the working distance decreases. **Working distance** is the distance between the specimen and the objective lens. Also the opening in each objective lens becomes smaller as you increase magnification. Many light rays will be refracted away from the lens and so the image may darken as you increase magnification. Adjust the light control or the iris diaphragm as needed. Notice also that the field of view (the area being viewed on the slide) decreases as magnification increases. Make sure before increasing magnification that the area to be viewed is close to the center of the field of view.

Oil immersion is used to view extremely small specimens such as bacteria. Bacteria must be viewed under oil immersion in order to determine true shape, size, and color. Different materials have a characteristic refractive index which is the amount light bends when passing through the material. When oil immersion is used, fewer light rays are refracted as they pass through the glass slide because glass and immersion oil have the same refractive index.

III. IDENTIFYING PARTS
OF THE COMPOUND LIGHT MICROSCOPE

Locate and **identify** the following parts of the microscope in **Figure 3.1**.

1. **Light Switch**—Located on the base of the microscope.
2. **Light Source** (Illuminator)—Provides the light that illuminates the specimen. This is located in the base of the microscope.
3. **Iris Diaphragm**—Enables the viewer to adjust the amount of light that reaches the specimen. This is located below the condenser. A short black lever opens or closes the diaphragm.
4. **Condenser**—Focuses or condenses the light. This is located underneath the hole in the stage.
5. **Condenser Control Knob**—Adjusts the height of the condenser. Use the metal knob, located in front of the coarse and fine adjustment knobs.
6. **Mechanical Stage**—A platform with a slide holder. Serves to hold slides in place.
7. **Stage Control Knobs**—Used to move the slide around on the stage. Located hanging underneath the stage. Note: there are two knobs that control movement in two directions.
8. **Objective Lenses**—Set of three or four lenses located directly above the stage. The magnification of each lens is engraved on the metal lens housing.
   a. **Scanning Power**—This lens magnifies 4X and is the shortest of the three lenses. It is used for initial focusing and viewing.
   b. **Low Power**—This lens magnifies 10X.
   c. **High Dry Power**—This lens magnifies 40X. No oil should be used this lens.
   d. **Oil Immersion Lens**—This lens magnifies 100X. Use only with the help of the instructor, since oil is needed.
9. **Revolving Nosepiece**—Allows the objective lenses to move into position above the specimen. Connects the lenses to the lower end of the body of the microscope.
10. **Body Tube**—A metal casing through which light passes to the oculars.
11. **Ocular (Eyepiece)**—This lens has a magnification of 10X, and is located in the uppermost part of the microscope. One of the oculars has a pointer that can be moved by turning the tube the ocular is housed within. If you have a binocular microscope, you can adjust the distance between the lenses to accommodate your eyes.
12. **Arm**—The upright structure attached to the base used in carrying the microscope.
13. **Base**—The heavy, flat support on which the microscope rests.
14. **Coarse Adjustment Knobs**—Used to rapidly alter the distance between the objective lens and the stage to focus on an object. These are the large knobs that extend from each side on the lower part of the arm. Use only scanning and low powered lens.
15. **Fine Adjustment Knobs**—Used to slowly alter the distance between the objective lenses and the stage when using the longer objective lenses (40X and 100X). These are the smaller pair of knobs that extend from each side of the lower arm.

IV. TOTAL MAGNIFICATION

The **total magnification** of the object can be calculated by multiplying the power of the ocular by the power of the objective.

Ocular	X	Objective	=	Magnification
_____	(Scanning)	_____		_____
_____	(Low)	_____		_____
_____	(High-dry)	_____		_____
_____	(Oil immersion)	_____		_____

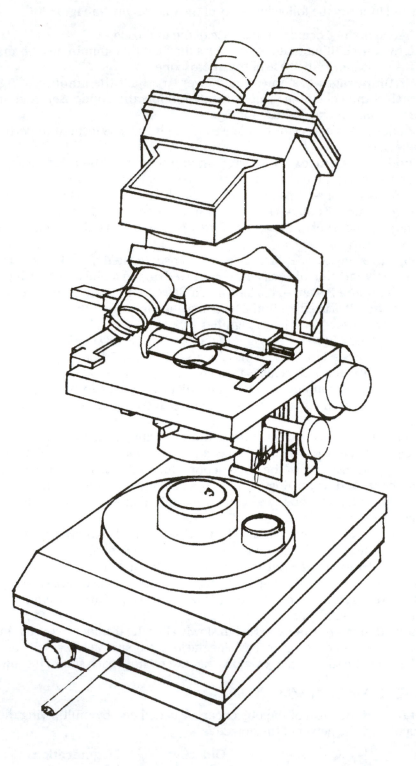

Figure 3.1   Schematic of a Binocular Microscope

V. USE AND CARE OF THE MICROSCOPE

1. Remove your assigned scope from the cabinet or shelf and using both hands carry it to your desk. The stage should face the user.
2. Check to see that the scope was put away properly by the previous user (see below) and report at once any scopes that were not.
3. Remove the cord carefully from the back of the stage and plug it in. Turn the light on.
4. Place the slide on the stage using the **mechanical stage clips.** Lower the 4X scanning objective to its lowest point. Check to see both oculars are equally adjusted.
5. When focusing on slides, always start with the 4X objective as close to the slide as it will go, and focus by turning the coarse adjustment away from the slide. Fine tune the image by using the fine adjustment. Use the iris diaphragm to adjust the amount of light to give the best resolution.
6. When switching to higher powered objectives, DO NOT touch the coarse adjustment. The microscope is **parfocal.** Only small changes in the fine adjustment and the amount of light are necessary.
7. ALWAYS turn back to 4X before removing a slide from the scope.
8. When you are finished with the scopes for the day, complete the following steps.
   a. Turn the 4X objective in place and raise it to its highest point or lower the stage to its lowest point.
   b. Turn off the light.
   c. Moisten lens paper with lens cleaner and wipe off the objective and ocular lens with it.
   d. Then wipe off the stage and the adjustment knobs with a piece of lens paper.
   e. Polish all of these with a dry piece of lens paper.
   f. Place the cord at the back of the stage.
   g. Show the microscope to your instructor before returning it to the proper cabinet or shelf.

VI. USING THE COMPOUND LIGHT MICROSCOPE*

1. Carry the microscope to the lab bench using **TWO HANDS. Never tilt the microscope, as this may cause the ocular lens to fall out.**
2. With the condenser control knob, move the condenser up to the hole in the stage. You should see the light shining though the condenser. Open and close the iris diaphragm using the protruding lever or roller bar in the condenser. Note the change in light intensity. Now open the iris diaphragm completely to allow for a maximum light.
3. Turn the objective lenses so that the scanning power lens (4X) is clicked into place.
4. Obtain a prepared slide of the letter "e". If necessary clean the slide—lens cleaner and lens paper. On the stage is a slide holder. Open the metal arm to make room for your slide. Gently place the slide into the holder, nestling the slide towards the back. Gently allow the metal arm to close on the slide. Make sure the letter e faces you as if you were reading it.
5. Practice moving the slide on the stage using the stage control knobs. Manipulate the slide so that the letter "e" is positioned directly over the condenser.
6. With the coarse adjustment knob, move the 4X objective all the way down, towards the slide. Now look through the ocular, and with the coarse adjustment knob, focus on the letter "e". (This is a good time to adjust the oculars for your eyes, if you have a binocular microscope.)

* From *Biological Investigations,* Revised Printing by Gayne Bablanian. Copyright © 2002 by Gayne Bablanian. Used with permission of Kendall/Hunt Publishing Company.

7. Move the "e" so that it is in the center of the field. Adjust the light intensity, if necessary.

   a. Move the slide to the LEFT. In which direction does the letter move? _____

   b. Move the slide to the RIGHT. In which direction does the letter move? _____

   c. Move the slide AWAY from you. In which direction does the letter move? _____

   d. Compare the orientation of the letter "e" when viewing it with the naked eye, then through the microscope. _____ _____

   e. Draw the letter "e" below.

Letter "e": Total Magnification _____ X

8. **WITHOUT TOUCHING THE COARSE FOCUSING KNOB,** turn the low power objective until it clicks into place. Using the coarse or fine adjustment knob, refocus on the letter. (Remember, the lenses are parfocal.) Draw the letter "e" below.

Letter "e": Total Magnification _____ X

9. Before going to the next lens, position the "e" so that part of the letter is directly in the center of your field of view. Now turn the high power (40X) objective until it clicks into place. Using the **FINE ADJUSTMENT KNOB ONLY,** refocus on the letter.

   a. Draw the letter "e" below.

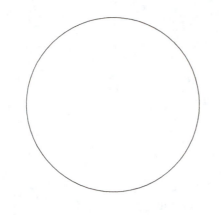

      Letter "e": Total Magnification _____ X

   b. What has happened to the field of view? _____

   c. What has happened to the letter "e"? _____

   d. What has happened to the light intensity? _____

   e. Why do you only use fine adjustment when viewing an object under high power? _____

VII. USING PREPARED SLIDES

Some of the slides you will be using in lab are prepared slides, like the letter "e" slides. Obtain the following slides, and draw a representation of each specimen in the space provided. You will have to determine the best magnification to use when drawing the object. Your instructor will help you to decide. If the slides are dirty use lens cleaner and lens paper to clean them.

1. *Daphnia* (Water Flea): Total Magnification _____ X

2. *Paramecium:* Total Magnification _____ X

3. Bacteria: Total Magnification _____ X

4. Leaf Cross Section (c.s.): Total Magnification _____ X

VIII. PREPARATION OF A WET MOUNT SLIDE

An alternative to using a prepared slide is to make your own slide, if a biological specimen is available. The **wet mount slide** is a temporary slide and can be used on living or preserved materials.

1. Obtain a clean glass slide.
2. Place your specimen (a small amount is generally the rule) onto the center of the **slide.** (see Figure 3.2)
3. Place the edge of a clean **coverslip** at the edge of the specimen. Slowly lower the coverslip over the specimen. This way, no air bubbles will get trapped.
4. When you are done viewing and drawing the specimen, discard the slide as instructed.
5. What is the purpose of using a coverslip?

_____

_____

_____

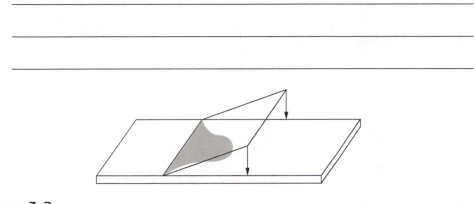

Figure 3.2

6. Make a wet mount slide and draw each of the following:

    a. *Elodea* leaf: Total Magnification _____ X

    b. hay infusion: Total magnification _____ X

REVIEW QUESTIONS

1. Define the following microscope terms:

    a. parfocal

    b. objective lens

    c. ocular lens

    d. iris diaphragm

    e. resolving power

    f. resolution

    g. numerical aperture

    h. total magnification

    i. working distance

2. Why is oil necessary for the best resolution of a compound light microscope?

3. What are the names given to the objective lenses and their powers?

4. Compare the high dry objective lens to the oil immersion objective lens.

5. Which objective should be in place when you first begin to focus?

6. On scanning power, which adjustment knob should be used for initial focusing?

7. Which knob is used to make fine details sharply visible?

8. Which is the only adjustment knob you use when looking through the high power objective?

9. Depending on your microscope, in which direction should you always focus, up or down?

10. With what should you always clean a lens?

11. The part of the microscope that controls the amount of light is called the _____.

12. If you are using a microscope having a 5X ocular and a 10X objective, what is the total magnification?

13. When you are finished with the compound microscope, which lens should be left in position?

From *Experiencing Biology: A Laboratory Manual for Introductory Biology,* Seventh Edition/ Revised Printing by GRCC-Biology 101 Staff, Biology Science Division. Copyright © 2002 by GRCC-Biology 101 Staff, Biology Science Division. Used with permission of Kendall/Hunt Publishing Company.

Exercise 4
The Cell

I. INTRODUCTION

All organisms are composed of the basic unit of life called the cell. Cells vary in their shape and size, yet share certain basic features in their design and in the way they function. Every cell contains a cell membrane, cytoplasm, ribosomes and nucleic acids. Each cell is a functional unit capable of carrying on all of the processes associated with life. As you compare one cell type to another, in today's lab, keep in mind that you will be observing cells at different levels of organization and complexity. You will be asked to compare the features of cells, as they relate to the life processes of organisms. (See Figure 4.1 for plant and animal cells. Figure 4.2 for bacterial cells.)

After completing this exercise, you should be able to:

1. List the differences between prokaryotic and eukaryotic cells.
2. Use the compound microscope to view cells and their structures.
3. Describe the characteristics of bacteria and cyanobacteria.
4. Describe and define the structures that you expect to see in animal and plant cells.

II. PROKARYOTIC CELLS

The simplest cellular organization is found in cells that belong to the **Kingdom Monera.** They may also be subdivided into the Domains Eubacteria and A. The **bacteria** and the **cyanobacteria** (blue-green algae) are said to be **prokaryotes** because they do not contain a membrane-bound nucleus or any other membrane-bound organelles. The DNA found in these cells is not organized into multiple chromosomes as in the cells of more complex organisms. These organisms also possess a cell wall.

A. Bacteria

Bacteria are examples of prokaryotic cells. They vary in size from 1–10 μm and have various shapes and arrangements (Figure 4.2). You will be observing living bacteria that are used in a commercial process that converts milk to yogurt. The species is called *Lactobacillus acidophilus* and is present in yogurt that contains **active cultures** of the organisms. Also, you will be comparing the sizes and shapes of other species of bacteria using prepared slides.

Procedure

1. Obtain a clean microscope slide and a coverslip. Place a drop of 0.9% NaCl water on the slide.
2. Using a toothpick, place a tiny dab of yogurt on the slide, and mix the yogurt with the drop of water.

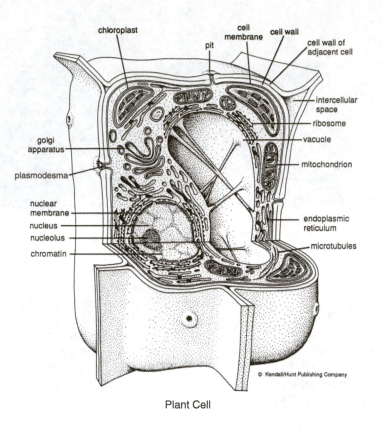

chloroplast

cell
membrane cell wall

pit

cell wall of
adjacent cell

intercellular
space

ribosome

vacuole

mitochondrion

golgi
apparatus

plasmodesma

nuclear
membrane

nucleus

nucleolus

chromatin

endoplasmic
reticulum

microtubules

© Kendall/Hunt Publishing Company

Plant Cell

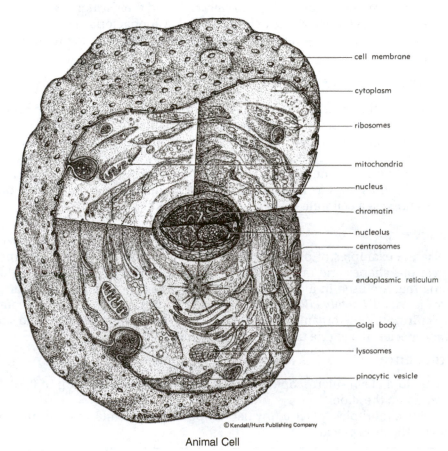

cell membrane

cytoplasm

ribosomes

mitochondria

nucleus

chromatin

nucleolus

centrosomes

endoplasmic reticulum

Golgi body

lysosomes

pinocytic vesicle

© Kendall/Hunt Publishing Company

Animal Cell

Figure 4.1

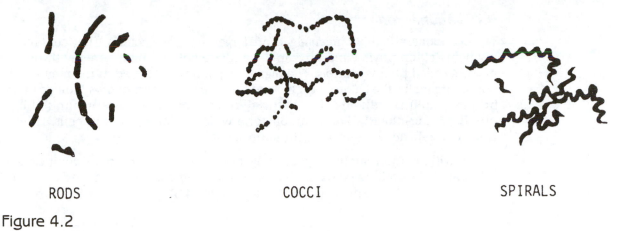

RODS   COCCI   SPIRALS

Figure 4.2

From *Laboratory Guide to Human Biology, Online Version* by Robert Amitrano and Martha Lowe. Kendall/Hunt Publishing Company.

3. Add a coverslip and examine the yogurt with a compound microscope.
4. Some secrets to success:
   a. Reduce the light, using the iris diaphragm, or drop the condenser to a lower spot.
   b. After initial focusing—the scanning objective lens. Focus with the low-power objective (1OX), then rotate the high-power objective into place, and refocus with the fine adjustment knob.
5. The bacteria appear as tiny rod-shaped structures, and are found between the "pieces" of yogurt.
6. Make a sketch of the cells below: Magnification—400X

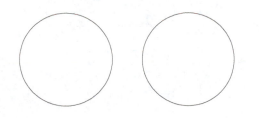

7. Now, obtain a prepared slide of different types of bacterial cells. Observe these using the microscope and draw them in the space provided below: Magnification—400X

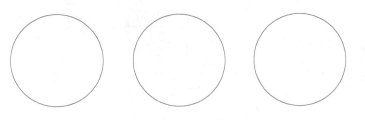

8. Using Figure 4.2 as a guide, list the different shapes of bacteria that you observed.

B. Cyanobacteria

The **cyanobacteria** (blue-green algae) are larger than the bacteria. They contain the pigments chlorophyll a and phycobilin, and are able to obtain energy through a process called photosynthesis. The pigments are not contained in the membrane-bound organelle, the chloroplast. Instead they are located on photosynthetic membranes called **thylakoids**. The cell wall of cyanobacteria is surrounded by a **mucilaginous sheath.** For cyanobacteria which form a group of cells called a **colony,** the sheath will surround the entire colony.

You will be observing two types of cyanobacteria. *Oscillatoria* is composed of cells formed into long chains. *Gleocapsa* is a green colony consisting of several cells surrounded by a mucilaginous sheath (Figure 4.3).

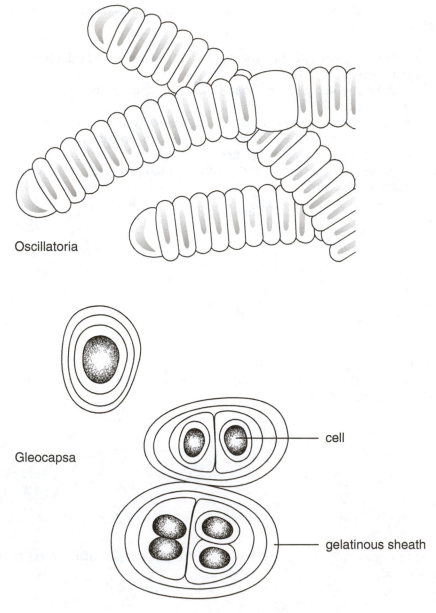

Oscillatoria

Gleocapsa

cell

gelatinous sheath

Figure 4.3

Procedure

1. Prepare a wet mount by placing a drop of water containing *Oscillatoria* on a clean slide and adding a coverslip.
2. Examine on high power, and draw a picture in the space provided below.
3. Repeat the above procedure with a sample of *Gleocapsa*.
4. If living material is not available, use the prepared slides provided by your instructor.
5. *Oscillatoria* (400X)   *Gleocapsa* (400X)

a. In which of the organisms, that you observed, is the mucilaginous sheath most prominent? _____
b. If you didn't know that these organisms were cyanobacteria, what would you have guessed they might be? _____
c. Why? _____

III. EUKARYOTIC CELLS

Eukaryotic cells are much larger than prokaryotic cells, and have a membrane-bound **nucleus** that contains chromosomes made of DNA organized around histone proteins. The cytoplasm contains membrane-bound organelles that are specialized structures each with a specialized function. Eukaryotic cells are found in the kingdoms **Protista, Fungi, Plantae** and **Animalia.** The 4 kingdoms are all placed into the domain Eukaryotae.

A. Protists

The protists include unicellular (one-celled) organisms and unicellular-colonial forms along with the multicellular algae. The algae are green and plant-like, while the protozoa are nongreen and animal-like. The protozoa *Paramecium* and *Ameba* will be used as examples of this kingdom (Figure 4.4).

Procedure

1. Make a wet mount of *Paramecium*. **NOTE:** These ciliated protozoa swim rapidly, so you must add a drop of methyl cellulose to the drop of culture fluid before adding the coverslip.

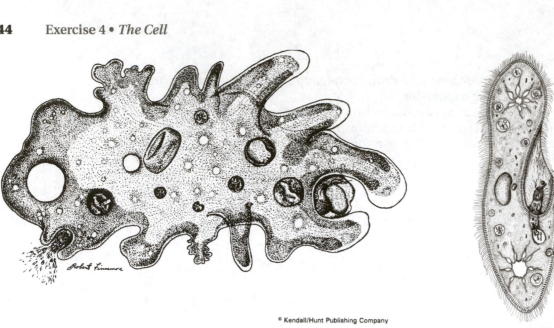

© Kendall/Hunt Publishing Company

AMEBA

© Kendall/Hunt Publishing Company

PARAMECIUM

Figure 4.4

2. Observe under high power and draw below: (400X) These organisms are alive so drop the light down so their water doesn't evaporate. Once the light is lowered look for the beating of the cilia.

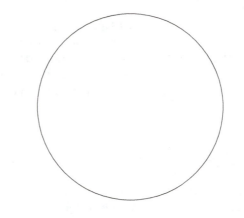

3. What is the shape of the *Paramecium?* _____

4. What extra features do you observe in this cell, which were not present in the cyanobacteria? _____
_____
_____

5. Make a wet mount of *Ameba*. **NOTE:** These organisms are located on the bottom of the jar. Do not mix the contents of the jar, or disrupt the bottom layer. Using the dropper, carefully draw a small amount of fluid from the bottom of the jar. (The fast moving organisms are not *Ameba*.)

6. Observe under high power and draw below: (400X) Note the movement of the cytoplasm of the ameba. The currents of cytoplasm push against the cell membrane allowing the ameba to move by cyclosis or cytoplasmic streaming.

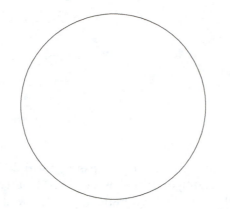

7. Describe how the *Ameba* is moving.

_____

_____

_____

8. Does *Ameba* have a cell wall? How can you tell?

_____

_____

_____

9. Compare the movement of *Ameba* to that of *Paramecium*.

_____

_____

_____

B. Fungi: (Optional)

Most of the members of the Kingdom Fungi are multicellular. The yeast, however, are unicellular. Using brewer's or baker's yeast, you can observe fungal cells and their internal structures. This type of yeast is used commercially in both the baking and brewing industry.

Procedure

1. Make a wet mount slide using a drop of yeast solution. NOTE: Add a drop of either cotton blue or methylene blue stain to the drop of yeast solution to help you view these cells.

2. Observe the cells under high power and draw below: (400X)

C. *Plants*

Plant cells are more complex and possess several additional cytoplasmic organelles. The outer covering of the cell is a rigid **cell wall** composed of **cellulose.** The cell wall overlies the plasma membrane. It protects and supports the cell while not interfering with the movement of molecules across the cell membrane.

In the cells of the pond plant, *Elodea,* you will observe colored organelles (usually green) called **chloroplasts** (Figure 4.5). Chloroplasts are the site for the process of photosynthesis. The nucleus in these cells will not be obvious because it is pushed to one side by the large, fluid-filled, **vacuole.** Be sure to look for **cyclosis** or cytoplasmic streaming. The chloroplasts may be moving around the inside of the cell membrane.

Also, you will be observing plant cell structure using the red onion. The onion is the root part of the plant. Therefore, photosynthesis does not occur in this portion of the plant. You will not find any chloroplasts in these cells.

Procedure

1. Obtain a sprig of *Elodea* and tear off a leaf from the top part of the sprig.
2. Make a wet mount of the leaf. If you place the top surface facing up, it will be easier to see the larger cells that are located on the top surface of the leaf.
3. Examine the leaf with your microscope. How many layers of the cells are present? _____
4. Under low or high power, focus on one cell. Each of the small, rectangular "boxes" represents a cell surrounded by the cell wall.
5. Chloroplasts appear as small green bodies within the cell. Observe the cell for at least a minute to see the phenomenon called cyclosis.
6. Draw the *Elodea* cell below:

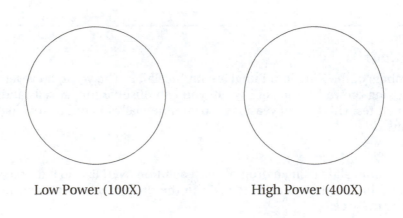

Low Power (100X)   High Power (400X)

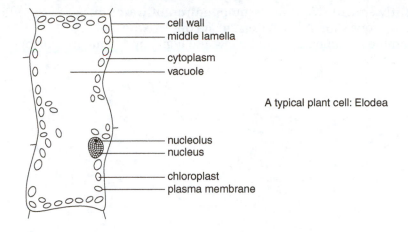

cell wall
middle lamella
cytoplasm
vacuole

A typical plant cell: Elodea

nucleolus
nucleus

chloroplast
plasma membrane

Figure 4.5

7. List the organelles and structures that are visible to you. (Hint: There should be at least four items in your list.)

   _____

   _____

8. Describe the process of cyclosis in your own words.

   _____

   _____

9. Obtain a piece of red onion. Snap the leaf backwards and remove the thin layer of epidermis formed at the break point (see Figure 4.6).

Selection of
Onion

SNAP BACKWARDS TO EXPOSE   PEEL OFF EPIDERMIS

Figure 4.6

10. Gently spread this thin tissue in a drop of water mixed with iodine on a microscope slide. Add a coverslip and observe the tissue.
11. Examine the onion cells under low and high power, and draw below.

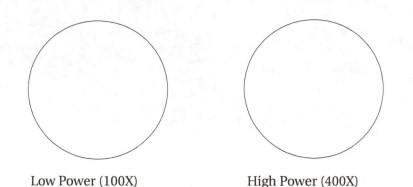

Low Power (100X)   High Power (400X)

12. List the organelles and structures that are visible to you. (Note: You should be able to see a round shaped body inside the nucleus. This is called the **nucleolus.)**

_____

_____

13. How do the two plant cells that you have examined differ in structure?

_____

_____

14. Since all plants have chloroplasts (somewhere in their overall structure), where are the chloroplasts located in the onion *plant?*

_____

_____

Optional Exercise: Potato Plastids

Plastids are organelles where food is manufactured and stored. The potato is an underground stem modified for the storage of starch. Within the cells of the potato you will examine **amyloplasts,** a type of plastid that stores starch.

Procedure

1. Using a sharp razor blade, make a very **THIN** section of the potato tuber. (The section should be thin enough to allow light to pass through.) Rinse the section with water.

2. Mount the slice with a drop of water on a slide. Add a coverslip.
3. Observe the potato cells. Within the cytoplasm are starch grains of various sizes that were produced in the amyloplasts.
4. Now add a drop of iodine to the edge of the coverslip. Draw the iodine under the coverslip by touching a small piece of paper towel to the opposite edge of the coverslip.
5. The details of the starch grains will now be more visible.
6. Iodine is a stain specific for starch. How do your potato results compare to the results you obtained with the onion?

_____

_____

7. What does this tell you about how carbohydrates are stored in the onion versus the potato?

_____

_____

D. Animals

Animal cells vary greatly in size, shape and function. Unlike plant cells, animal cells **do not** have a cell wall, chloroplasts, or vacuoles. The plasma membrane protects the cell, and allows for more flexibility in the overall structure of the animal body. A readily available type of animal cell is the human epithelial cell found inside the mouth, on the inner surface of the cheek (Figure 4.7).

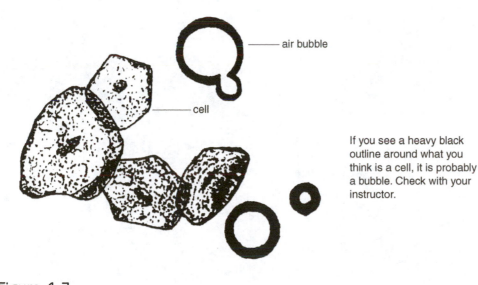

air bubble

cell

If you see a heavy black outline around what you think is a cell, it is probably a bubble. Check with your instructor.

Figure 4.7

Procedure

1. Place a small drop of water on a microscope slide.
2. Using the flat edge of a clean toothpick, gently scrape the inner surface of your cheek.
3. Mix the "scraping" into the water. Discard the toothpick into the RED BIO-HAZARD CONTAINER.
4. Using a toothpick add a small drop of methylene blue stain and add a coverslip.
5. The cheek cells are flat and sometimes are seen in clusters. Under low power, search for bluish cells with small darker blue areas in the center.
6. Switch to high power and draw several cells below: (400X). Label the nucleus, plasma membrane, and the cytoplasm.

High power (400X)

E. The Plasma Membrane

One common feature in all cells you observed to this point was the cell membrane. Even the cells which had a cell wall had a membrane immediately inside the wall. All cells have this membrane and it functions in controlling what enters and leaves the cell. You have already observed one of these methods. When you exposed the *Elodea* cells to the salt solution water moved out of the cell. This process is known as **osmosis,** a special kind of diffusion involving only water. In osmosis, water moves across the membrane from an area of high concentration to an area of low concentration. When the salt solution surrounded the *Elodea* cells, the concentration of water was not the same on both sides. Water inside the cell was about 99% pure while water on the outside of the cell was 90% pure. Osmosis occurred and water left the cell (from the higher concentration inside the cell to the lower concentration outside).

To enable you to understand osmosis better, you and your lab partners should get three slides for your lab table. On the first slide place a drop of distilled water, on the second slide place a drop of 10% NaCl solution and on the third slide place a drop of 0.9% NaCl solution. Using a clean toothpick for each slide, sheep blood needs to be mixed into the water and solution drops. Put the toothpick into the sheep blood and then mix it gently into the drop on the slide. Throw the toothpick into an orange biohazard bag and cover the mixture with a coverslip. Observe each slide with high power of the microscope. Compare the size, shape and condition of the cell membrane for each of the conditions.

	Size	*Shape*	*Sketch of Membrane*
Slide 1–H$_2$O			
Slide 2–10% NaCl			
Slide 3–0.9% NaCl			

IV. COMPARISON OF BACTERIAL, ANIMAL, AND PLANT CELLS

In the table below, indicate which structures and organelles are PRESENT or ABSENT for each category of organism. Use your textbook or any other information to complete this exercise.

	Bacterium	*Plant*	*Animal*
Cell Wall			
Plasma Membrane			
Flagella			
ER			
Ribosomes			
Golgi Apparatus			
Nucleus			
Chloroplasts			
Mitochondria			
Chromosomes			
Vacuoles			

V. QUESTIONS

1. Which is larger, *Lactobacillus* or the cyanobacteria *Oscillatoria?*

2. Where is the chlorophyll located in the *Gleocapsa* cell?

3. Can you see any internal structures in a bacteria or cyanobacteria cell?

4. Does the appearance of a bacteria and cyanobacteria cell indicate that they are primitive or highly developed? Explain your answer.

5. Does the *Ameba* have a ridged distinct shape?

6. Does the *Ameba* have a cell wall?

7. Does the *Paramecium* have a characteristic shape?

8. Does the *Paramecium* have a cell wall?

9. Compare the difference in movement between the *Ameba* and *Paramecium.*

From *Laboratory Manual for General Biology, Third Edition,* Delgado, C. C. Kendall/Hunt Publishing Company.

10. What cellular structure/structures in the onion epithelial cells are stained by iodine?

11. Describe the shape of the onion cells.

12. While viewing *Elodea* cells did you observe cytoplasmic streaming? What is its purpose?

13. What are the numerous, small, round, green structures in the *Elodea* leaf cells?

14. What structures that you see indicate that the *Elodea* cells are plant cells?

15. In the human cheek cells, what are the round, darkly stained structures?

16. Do animal cells have a cell wall?

VI. FURTHER EXERCISES

In the table that follows, give a description of each eukaryotic cell structure and its function in the cell.

From *Laboratory Manual for General Biology, Third Edition*, Delgado, C. C. Kendall/Hunt Publishing Company.

Structure	Description	Function
Cell Wall		
Plasma Membrane		
Flagella		
Cytoskeleton		
ER		
Ribosomes		
Golgi Apparatus		
Nucleus		
Nucleolus		
Chromosomes		
Mitochondria		
Chloroplasts		
Lysosomes		

Cell Reproduction

I. INTRODUCTION

Mitosis and meiosis are two mechanisms by which the nuclei of cells divide. In **mitosis,** the parental cell divides producing **two daughter cells.** The daughter nuclei are **identical** to each other and to the parental nucleus in chromosome number and genetic makeup. The number of chromosomes per cell varies from one species to another. Human cells have 46 chromosomes (**23 homologous pairs**). **Homologous chromosomes** look alike, but carry different forms of the same genes. In higher plants and animals, having two sets of homologous chromosomes is called the **diploid** condition and is designated by the symbol **2N.**

Meiosis, in contrast, is a part of the **sexual life cycle** in which the daughter nuclei produced are found in cells that differentiate as male and female gametes: the **sperm** and **eggs.** Meiosis results in **four daughter cells** each containing **one-half** the chromosome number (**haploid** or N) and **different genetic composition.**

II. THE CELL CYCLE

The complex series of events that encompasses the life span of an actively dividing cell is called the **cell cycle** (Figure 5.1). After a cell is produced, it enters a period of growth and development termed **interphase** during which it grows to its maximum size and duplicates its chromosomes in preparation for cell division. During interphase the following events occur:

1. **G1** (gap 1) **phase**—Occurs after mitosis and cytokinesis, and is the primary growth phase of the cell.
2. **S** (synthesis) **phase**—DNA replication occurs. This is necessary so that each new cell will contain a complete copy of the genetic material.
3. **G2** (gap 2) **phase**—The second growth phase, in which preparations are made for mitosis. Mitochondria and other organelles replicate, chromosomes condense, and microtubules begin to assemble at a spindle.
4. **Cell division** occurs at the end of the cell's life cycle. **Nuclear division** is called **mitosis,** and is divided into 4 subphases: **Prophase, Metaphase, Anaphase,** and **Telophase.** Division of the cytoplasm into two daughter cells is called **cytokinesis.**

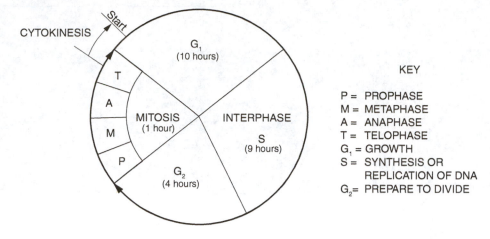

Figure 5.1 The Cell Cycle

III. MITOSIS IN PLANT CELLS

The Alium (onion) root tip is an excellent model to use to study the events of mitosis. It contains an area called a **root meristem,** where there is active cell division. Within this region, you will be able to find every stage of mitosis.

Procedure

1. Obtain a prepared slide of a longitudinal section through an onion root tip. Search for the stages of mitosis, as outlined below. Draw each stage in the spaces provided.

2. **Interphase**
   A cell in interphase is "between divisions." The cytoplasm appears granular and a nuclear membrane is present. In these preserved and stained cells, the genetic material within the nucleus usually appears in the form of scattered CHROMATIN GRANULES. In the living state, fine filaments can be found which become thicker as cell division approaches. They are then known as CHROMOSOMES. Duplication of chromosomes occurs during interphase. The pairs of (sister) CHROMATIDS that result are not visible in the slide preparation you are studying. They remain joined together at a point called the CENTROMERE.

3. **Mitosis Stages in the Onion Root Meristem**
   Mitosis is a continuous process which is completed in a relatively short time. For convenience, the process is divided into stages.

   As you study the slide, it is not necessary to find the stages in order. However, it is important that they appear in order on your drawing page. If you happen to see a good telophase while looking for some other stage, stop, study, and draw it in the correct space on the drawing sheet provided. Use posters and models to help you identify the stages.

   a. **Prophase**
      The nuclear membrane and nucleolus have dissolved but distinct chromosomes, each composed of paired chromatids, are now visible in the space formerly occupied by the nucleus. The onion has 16 chromosomes (now 32 sister chromatids) but counting them is impossible.

DRAW one cell in prophase and label: chromosomes and cell wall.

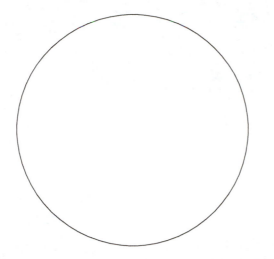

b. **Metaphase**

A fully-formed mitoic spindle now exists with fibers extending from one end of the cell to the other. These may not be visible. The point near each end of the cell where the spindle fibers converge is a spindle POLE (the term "pole" is a figure of speech like "north pole" or "south pole"). The sister chromatids are grouped in the center on an EQUATORIAL PLANE. SPINDLE FIBERS connect the CENTROMERES with the poles, however neither are visible with our microscopes.

DRAW a metaphase cell and label the region of the equatorial plane and the chromosomes.

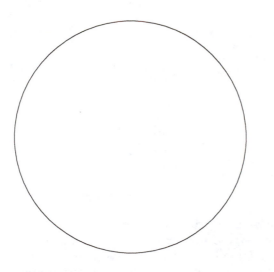

c. **Anaphase**

In this stage, the centromeres joining each pair of chromatids split and chromatids become chromosomes. The chromosomes are moved along spindle fibers toward opposite poles. Each chromosome bends at its centromere assuming the shape of a letter "V" as it is moved through the cytoplasm.

DRAW and label a cell in anaphase: chromosomes and poles.

d. **Telophase**
Having reached the poles, each group of chromosomes condenses into a mass in which the individual chromosomes are no longer visible.

Division of the cytoplasm (cytokinesis) begins in telophase. A **CELL PLATE** or new cell wall forms in the center of the cell at right angles to the spindle axis.

DRAW one cell and label: cell plate and chromosomes.

4. **Cytokinesis: Dividing the Cytoplasm**
Plant cells undergo cytokinesis by forming a cell plate across the equator of the cell. A cell plate is a collection of vesicles formed by the Golgi apparatus that contain cellulose and other cell wall components. The vesicles fuse to form plasma membranes and a cell wall that extends across the midplane of the cell. This divides the cytoplasm in half to form two daughter cells. Cytokinesis begins in telophase and is completed in interphase.

Find a pair of small cells in which the new cell wall has just been completed. These are new daughter cells.

From *Experiencing Biology: A Laboratory Manual for Introductory Biology,* Seventh Edition/Revised Printing by GRCC-Biology 101 Staff, Biological Science Division. Copyright © 2002 by GRCC-Biology 101 Staff, Biological Science Division. Used with permission of Kendall/Hunt Publishing Company.

DRAW a pair of daughter cells and label: new cell wall, old cell wall, nucleus, and nucleous.

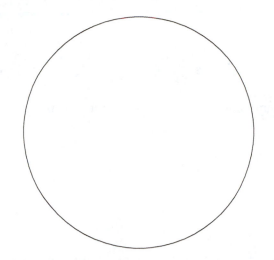

IV. MITOSIS IN ANIMAL CELLS

Animal cells differ from plant cells in several ways, the most evident being that animal cells lack a cell wall. Unlike plant cells, animal cells have two pairs of **centrioles** that form the center of each pole during mitosis. The arrays of microtubules that completely surround each pole in an animal cell are called **asters.**

In animal cells, **cytokinesis** involves the formation of a **cleavage furrow.** A contractile ring of microfilaments pinches the cell in half to form two daughter cells. The contractile ring is similar to placing a string around a balloon and pulling it tightly to form two balloons.

You will be observing mitosis in cross sections through the blastula (a stage of development) of the whitefish.

Procedure

1. Obtain a prepared slide of whitefish blastula cross sections.
2. Under high power, identify the stages of mitosis, and draw each phase in the space provided.
3. Label the following structures in your drawings. Chromosomes, spindle fibers, asters, cleavage furrow and cell membrane.

ANIMAL (whitefish blastula)*

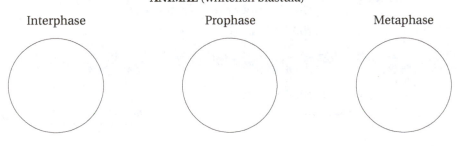

| Interphase | Prophase | Metaphase |

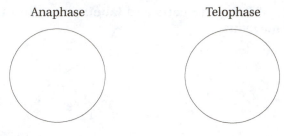

Anaphase   Telophase

DRAW one cell and label: cell membrane, nucleus, nucleolus, chromosomes and cytokinesis.

V. ESTIMATING THE DURATION OF THE PHASES OF THE CELL CYCLE

Cells from the apical meristem of the onion root tip have a cell cycle that lasts approximately 20 hours. By counting the number of cells on your slide in each phase of the cell cycle and doing some calculations, you can estimate the number of hours that a cell spends in each phase of the cell cycle.

1. Return the Alium root tip slide to your microscope. Locate the apical meristem region of one of the root tips and study it at H.P. Count all of the cells in one H.P. field that are in interphase. Record the number in the table below and on the board.
2. Continue examining the same H.P. field of view. Count all of the cells in each of the 4 subphases of mitosis. Record these numbers in the table below and on the board.
3. If you have not counted at least 100 cells total, select another high power field at random and repeat the above steps. The more cells you count, the more accurate your estimates will be.
4. Record the total number of cells counted beneath the table.
5. Use the following equation to determine the duration (in hours) of each phase of the cell cycle.

$$\text{Duration} = \frac{\text{number of cells in a particular phase}}{\text{total number of cells counted}} \times 20 \text{ hours}$$

6. Record your results in the table below.

Phase of Cell Cycle	Number of Cells Counted in Each Phase	Duration in Hours
Interphase		
Prophase		
Metaphase		
Anaphase		
Telophase		

Total number of cells counted _____

Which phase of the cell cycle lasts the longest? _____

Which subphase of mitosis lasts the longest? _____

How do your results compare with published results? _____

_____

How could you improve the accuracy of your estimate? _____

_____

VI. MEIOSIS*

Unlike mitosis, which is used for growth and repair, meiosis is a reduction cell division. **Meiosis** produces male and female gametes (sperm and eggs) that are needed for sexual reproduction. Meiosis is a sequence of two nuclear divisions that reduces the chromosome number by half. During fertilization (union of gametes), the diploid number is restored and a diploid (2N) zygote is produced. Diploid means there are two chromosomes of each type in a cell. Meiosis reduces the diploid number (2N) by half, to the haploid number (N). When one diploid cell proceeds through the two nuclear divisions of meiosis, four haploid cells will be produced.

Meiosis differs in several respects from mitosis. Meiosis consists of two nuclear divisions called meiosis I and meiosis II. Each of these divisions has a prophase, metaphase, anaphase and telophase (Figure 5.4). There is no chromosome duplication between the two divisions. The four haploid daughter cells are genetically different from each other as a result of events that take place during meiosis I.

In prophase of meiosis I, the homologous chromosomes are physically paired. This process is called **synapsis.** During synapsis, genes can be swapped between the homologous chromosomes that are adjacent to each other in a process referred to as **crossing over** (Figure 5.2).

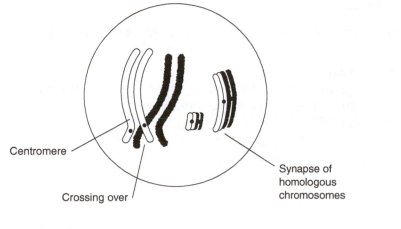

Centromere

Crossing over

Synapse of homologous chromosomes

Figure 5.2

_____

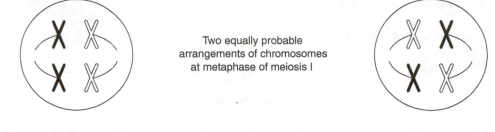

Two equally probable
arrangements of chromosomes
at metaphase of meiosis I

Figure 5.3

Also, another point at which genetic variation can occur is during metaphase of meiosis I. Pairs of homologous chromosomes become aligned on the equator independently of the other pairs. The number of chromosome arrangements possible depends on the number of homologous pairs of chromosomes present. For example, if a cell has two homologous pairs of chromosomes, two equally likely arrangements are possible (Figure 5.3). When the homologous chromosomes separate in anaphase I, the two daughter cells have different combinations of maternal and paternal genes. This source of genetic difference is called **independent assortment.**

Together, independent assortment and crossing over ensure that no two gametes of a single individual will be genetically identical. These processes, coupled with the process of fertilization, further ensure the genetic variation in a population of organisms. Genetic variation increases the chances of survival in changing environments.

A. The Events of Meiosis

In the flowering plants, meiosis occurs in the ovary and anthers of the flower. You will be observing the stages of meiosis in the **lily** (**Lilium**). After the description of each phase of meiosis, find the stage in the representative **prepared slides.**

Procedure

Following each phase of meiosis, draw a representative cell from the prepared slides of *Lilium*.

1. **Interphase**
   a. Chromosomes become duplicated (each chromosome will have two chromatids joined at the centromere).
   b. Centriole duplication begins.

2. **Meiosis I**
   a. **Prophase I**
      1. chromosomes become visible
      2. synapsis occurs (homologous chromosomes pair up)
      3. crossing over occurs (parts of adjacent nonsister chromatids exchange places)
      4. centrioles move towards opposite poles (animal cells only)
      5. the spindle begins forming
      6. the nuclear envelope starts breaking up during the transition to metaphase

   b. **Metaphase I: homologous pairs align at the spindle equator**

   c. **Anaphase I**
      1. sister chromatids remain attached at their centromeres and move as a unit towards the same pole
      2. homologous chromosomes move towards opposite poles

   d. **Telophase I**
      1. chromosomes reach the poles
      2. a nuclear envelope forms around the chromosomes at each pole
      3. each pole has a haploid number of chromosomes that are all duplicated

3. **Cytokinesis occurs (division of the cytoplasm)**

4. Meiosis II

   a. Prophase II

      1. centrioles move towards opposite poles (animal cells only)

      2. the spindle begins forming

      3. the nuclear envelope starts breaking up during the transition to metaphase II

   b. Metaphase II: all chromosomes line up halfway between the poles

   c. Anaphase II

      1. sister chromatids separate

      2. once-joined sister chromatids are now called daughter chromosomes

      3. the daughter chromosomes move centromere-first to the poles of the cell

   d. Telophase II

      1. chromosomes reach the poles

      2. a nuclear envelope forms around the chromosomes at each pole

      3. the spindle disappears

      4. each pole has a haploid number of chromosomes that are all unduplicated

5. Cytokinesis occurs (division of the cytoplasm)

B. Gametogensis

The formation of **gametes** (the results of meiosis) is called **gametogenesis.** Cellular maturation is required before newly formed cells from meiosis become functional gametes. In mammals, the formation of sperm cells is called **spermatogenesis,** and results in **four sperm** cells being produced. The formation of egg cells is called **oogenesis,** and results in **one egg cell** being produced (see Figure 5.4).

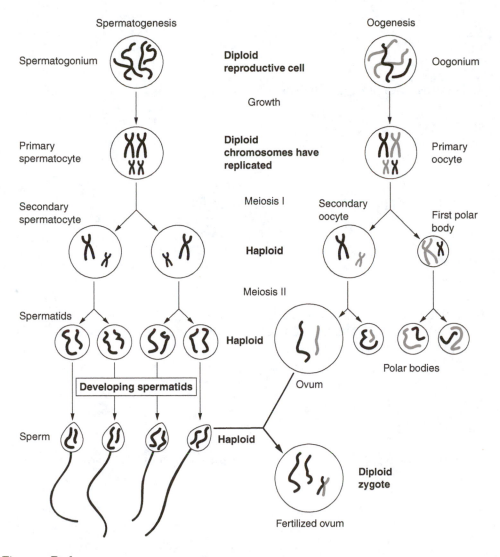

Figure 5.4

VII. QUESTIONS

1. How many chromatids are found in each chromosome during G2 phase?

2. How many chromosomes are maternal in a human cell with forty-six chromosomes?

3. Are chromosomes visible during most of the cell cycle?

4. Is mitosis the same as cell division?

5. Explain why plant cells can't form cleavage furrows.

6. A cell with forty-six chromosomes goes through one round of mitosis and cytokinesis. How many chromosomes are inside the nucleus of each of the two resulting cells?

7. During which phase of the cell cycle does the amount of DNA in the cell double?

From *Laboratory Manual for General Biology, Third Edition,* Delgado, C. C. Kendall/Hunt Publishing Company.

8. During which phase of mitosis does a chromatid become a chromosome?

9. How many chromosomes are found in human sperm and egg cells?

10. Explain the differences between the movements of chromosomes in anaphase I and anaphase II.

11. Explain how crossing over increases the variety of gametes produced during meiosis.

12. Describe the similarities and differences between mitosis and meiosis.

13. What is the purpose of cell division?

14. Where in the human body would you likely encounter high rates of cell division?

From *Laboratory Manual for General Biology, Third Edition,* Delgado, C. C. Kendall/Hunt Publishing Company.

Genetics

I. INTRODUCTION TO HUMAN GENETICS

To butcher a well-known phrase, "what you see is what you got." This is a simple way of describing the principle of the **phenotype.** Most cells in the human body have 23 pairs of **chromosomes.** One set of 23 chromosomes was originally contributed by each parent. Chromosomes are subdivided into **genes** or segments of DNA that code for proteins. For each hereditary trait there are two or more genes called **alleles** (variations of a given gene) which contribute to its expression. A single chromosome contains one allele for each gene. Each parent contributes one gene. The genetic (or allelic) combination that gives rise to a hereditary trait is called the **genotype.** The physical expression of the trait is called the **phenotype. Dominant traits** result from either of two genotypes: **homozygous dominant** (EE, two dominant alleles) or **heterozygous** (one dominant and one recessive allele, Ee). **Recessive traits** are an expression of a **homozygous recessive** genotype (two recessive *alleles,* ee).

In this laboratory exercise, you will be doing a great deal of "self-analysis." You, with the help of your lab partner, will determine which set of hereditary traits you possess. All the students will be polled to obtain a phenotypic characterization of the class.

II. HEREDITARY TRAITS

Procedure

For each of the sixteen traits discussed below characterize your individual phenotype as either dominant or recessive. Circle your genotype under "personal data" on the data table provided in the results and questions section. Note that **capital letters** indicate **dominant** alleles while **lower case letters** indicate **recessive** alleles.

Eye Color

1) If you have dark eyes (black, brown or hazel green), then they are dominant over the recessive light colors of blue and gray.
    A = dark eyes (dominant)
    a = light eyes (recessive)

Eyelashes

2) Measure the length of your longest eyelashes with a metric ruler. You may need the help of your lab partner to do this. Use the following standards to determine the dominant or recessive state of your lashes.

    1 cm or longer = long
    less than 1 cm = short

    Long eyelashes are dominant over short eyelashes.
    B = long eyelashes (dominant)
    b = short eyelashes (recessive)

Hair

3) Several characteristics will be considered here. The colors considered to be dark or brunette are dominant over light or blond hair.
    H = dark hair (black, brown or dark red)
    h = light hair (blond or light red)

4) Non-red hair is dominant over red hair.
    D = non-red hair
    d = red hair

5) Examine your hairline at the forehead. If the hairline comes down to a point at the center of the forehead, this is called a "widow's peak" and it is dominant over a straight hairline.
    E = widow's peak
    e = straight hairline

Facial Characteristics

6) Do you have dimples? If you do, then that trait is dominant over a non-dimpled trait. Your dimples may be single (one cheek) or on both cheeks.
    F = dimples
    f = no dimples

7) Freckles are dominant over no freckles.
    G = freckles
    g = no freckles

8) Have your lab partner check your ears. (Not for wax!) Are your ear lobes attached or are they free from the side of the head?
Free ear lobes are dominant over attached ear lobes.
    H = free ear lobes
    h = attached ear lobes

9) Do you have a Roman nose? (Not roamin' all over your face!) Does your nose have a distinct bend in the middle or is it straight? A Roman nose or bent nose is dominant over a straight nose.
    I = Roman nose
    i = straight nose

10) Full, thick lips are dominant over thin lips.
    J = full, thick lips
    j = thin lips

Taste

11) Obtain a small piece of PTC paper and chew it. Do you notice a bitter taste or only the taste of paper? The ability to taste the bitter substance PTC is a dominant characteristic over the lack of ability to taste PTC. Approximately 70% of the U.S. Caucasian population are tasters, and about 90% of the U.S. African American population discern the taste of PTC.
    K = ability to taste PTC
    k = inability to taste PTC

Feet and Hands

12) Take off your shoes. (Breath holding is optional at this point in the lab.) Is your second toe longer than your big toe? If so, this is a dominant trait over having a shorter second toe.
    L = second toe equal to or longer than big toe
    l = second toe shorter than big toe

13) Place both hands in front of your face with the palms toward you. Place your little fingers side-by-side. Do they run parallel to each other for the entire length or do the last sections of the fingers turn out away from each other? Bent little fingers are dominant over straight little fingers.
    M = bent little fingers
    m = straight little fingers

14) What hand are you using to record data? Do you use your right hand or your left hand? Right-handedness is dominant over left-handedness, although social bias can greatly influence the expression of this trait.
    N = right-handed
    n = left-handed

Height

15) More than likely, short stature is dominant over tall stature. If you are male, tallness is defined as being over 178 cm (5'10") and for females as being over 170 cm (5'7").
    P = short stature
    p = tall stature

Tongue

16) Take out some frustrations on your lab partner or your instructor. Stick out your tongue and try to roll it into a longitudinal "U-shaped" trough. This ability is dominant over the lack of ability to roll the tongue.
    Q = tongue roller
    q = not a tongue roller

Results and Questions

1) Upon conclusion of your "self-analysis," enter your data on the board. Record the class data in the space provided in the data table below. Using these data calculate the occurrence of each trait (% dominant and % recessive) in your class population.

   Example:

   eye color   dark   15
                light   5
   total in class (n)   20

   dominant: 15/20 = 0.75 × 100 = 75%
   recessive: 5/20 = 0.25 × 100 = 25%

Table 6.1 Phenotypes

	Personal Data Circle your information		Class Data class size(n) = _____			
Trait	Dominant	Recessive	# Dom	% Dom	# Rec	% Rec
eye color	A-	aa				
eyelashes	B-	bb				
hair color	C-	cc				
red hair	D-	dd				
hairline	E-	ee				
dimples	F-	ff				
freckles	G-	gg				
earlobes	H-	hh				
nose	I-	ii				
lips	J-	jj				
PTC	K-	kk				
toe	L-	ll				
fingers	M-	mm				
handedness	N-	nn				
stature	P-	pp				
tongue	Q-	qq				

2) Which was the most common (highest percentage) dominant trait in the class?

_____

_____

3) What was the rarest dominant trait? _____

4) Does the fact that a trait is "dominant" imply anything about how common or rare it is in the class population? _____ Try to explain: _____

_____

5) Why are the dominant genotypes all written as a capital letter followed by a dash (e.g., B-)? _____

_____

6) Many of the traits examined in this lab are subject to social bias and/or environmental influence. Identify one such trait and discuss: _____

_____

7) Construct a histogram (bar-graph) on the graph paper provided using the class data. Indicate traits on the X-axis and the percent of the class *recessive* for that trait on the Y-axis. Label the axes and provide a descriptive title.

8) Summarize the findings from your graph. (Remember, a graph summary describes any trends or correlation in the data.) _____

_____

Procedures: (Work in pairs)*
Mendel's Law of Segregation and Chi Square

Gregor Mendel (1822–1884) was an Austrian monk who is generally credited with being the father of genetics. He was the first to discover the basic mechanisms involved in heredity. He formulated his findings in two statements which are commonly referred to as **Mendel's Law of Inheritance.** Mendel's first law is the Law of Segregation and is stated as follows:

The units of inheritance (genes) exist in pairs in an individual. In the formation of gametes, these genes separate, one to each gamete, so that each gamete has only one of each kind of gene.

A trait can occur in two or more different forms. Therefore, a gene is composed of **alleles** that govern the expression of that trait. According to the Law of Segregation, these alleles are separated from each other into different gametes in the cell division process known as **meiosis.** As a result of meiosis, the individual would be able to produce **two** different **kinds** of gametes as far as the given characteristics are concerned: one gamete containing one allele, the other gamete containing the other allele. If two individuals, each having contrasting alleles for a

From *Human Biology Laboratory Manual,* Fourth Edition/Revised Printing by Keith Cunningham and Leslie Snider. Copyright © 2001 by Kendall/Hunt Publishing Company. Used with permission.

*From *Human Biology: A Laboratory Manual,* Fifth Edition/Revised Printing by Roberta B. Williams. Copyright © 2000 by Kendall/Hunt Publishing Company. Used with permission.

given characteristic (heterozygous condition), are mated, it is possible to predict the ratio of the genotypes that will be produced. For example, in mice there is a gene that causes brown fur and a gene that causes white fur. The gene for brown is dominant over the gene for white. A mouse heterozygous for fur color would have a **genotype** consisting of an allele for brown (B) and an allele for white (b). If a monohybrid cross (only **one** trait involved) is made between two mice heterozygous for fur color, the possible genotypes of the off-spring could be determined with the following steps:

According to Mendel's First Law, each of these genes would segregate (separate) from one another into a gamete,

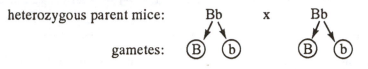

The Punnett Square Technique can be used to determine the possible types of genotypes (types of genes present) that the offspring might have:

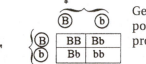

Genotypes in square are the possible zygotes that can be produced.

The possible zygotes are: BB, Bb and bb. Since there is one BB, two Bb, and one bb, the ratio among these offspring would be 1:2:1. To determine if this ratio would in reality be achieved, the following procedures should be followed:

1. Obtain a container of mixed black and white beans. Be sure there are 40 black beans and 40 white beans in the jar. This represents the potential gametes that a heterozygous (Bb) mouse can produce.
2. Obtain another contained of mixed black and white beans. Again, be sure there are 40 black and 40 white beans in the jar. This represents the potential gametes of a second heterozygous mouse (Bb).
3. Thoroughly stir the beans in each jar.
4. Obtain three paper cups. These are to hold the possible types of zygotes these two mice can produce. 1 cup should hold only black beans, 1 cup should hold only white beans and the third cup will be for a mixture of black and white beans.
5. Without looking, reach into a jar (parent mouse) and remove a bean (this would represent a gamete).
6. Randomly reach into the other jar and remove a bean (gamete). Fertilization now occurs. You have produced a zygote. What is its genotype? What color will it be (black is dominant over white)?
7. Randomly reach into each parent jar again, removing a bean (gamete) from each jar without looking. If these two beans (gametes) are identical in color place them in the same zygote cup. If they are a different combination of bean colors, put them in a different zygote cup.
8. Continue removing beans, two at a time (one from each parent), until all the beans have been paired in the three cups. One cup will contain BB (black, black), the second Bb (black, white), and the third bb (white, white) zygotes.
9. Count the number of zygotes in each cup (the number of zygotes in a cup can be found by counting the beans in the cup and dividing by 2).
10. What is the genotypic ratio (BB:Bb:bb) among the zygotes produced by the parent heterozygous mice? _____
11. You probably did not come out with an exact 1:2:1 (20:40:20) ratio. If not, it is possible to determine if your results are significantly similar to the "ideal" ratio of 1:2:1 by using a statistical technique known as the **Chi Square Test for Significance.**

Table 6.2			
	BB	*Bb*	*bb*
Expected number of zygotes (e)	20	40	20
Actual number of zygotes			
Deviation (d) = difference between expected and actual			
Deviation squared (d2)			
d2 divided by e			

12. Without explaining the mathematical derivations, the following procedure can be used to determine if your results are close enough to the expected ratio (1:2: 1) to indicate that **Mendel's Law of Segregation** is probably correct. Remember: total zygotes = 80; expected ratio among the 80 = 20:40:20 (which is a 1:2:1 ratio).
    Chi square value = sum of all the numbers in bottom row

    Chi square = _____

    Chi square value of 5.990 or less, deviation considered insignificant.

    Chi square value of 5.991 or greater, deviation considered significant.

13. If your Chi square value is insignificant, then your genotypic ratio among the zygotes is not significantly different from the ratio that Mendel said would result if his Law of Segregation is correct.
14. If your Chi square is significant, either Mendel's Law of Segregation is not correct or you did something wrong.
15. Carefully replace 40 black beans and 40 white beans in each of your parent bean jars.

Application and Perspective*

Punnett squares are a convenient device used in determining potential allelic combinations in offspring produced by sexual reproduction. From the 16 pheno-types you observed in class, choose one that is variable within your immediate family (for example, some left-handed and some right-handed members). Construct a Punnett square for the inheritance of this trait for each allelic possibility. Describe the results from your Punnett square. Further deduce your genotype knowing the phenotype(s) of grandparents and siblings if you can. Calculate a ballpark estimate for the likelihood of receiving your phenotype.

From *Human Biology: A Laboratory Manual*, Fifth Edition/Revised Printing by Roberta B. Williams. Copyright © 2000 by Kendall/Hunt Publishing Company. Used with permission.

*From *Human Biology Laboratory Manual*, Fourth Edition/Revised Printing by Keith Cunningham and Leslie Snider. Copyright © 2001 by Kendall/Hunt Publishing Company. Used with permission.

Pedigrees

In standard pedigree nomenclature, a circle indicates a female and a square represents a male. A horizontal line between them indicates a mating and the offspring are suspended below by a vertical line. If the circle or square is crosshatched (shaded), the individual is "afflicted" with whatever trait is being considered (that is, the individual expressed that phenotype). If the circle or square is left open (unshaded), then the individual does not express the trait, but still could be a carrier.

From the information and symbols given for each pedigree, you should be able to determine the genotypes for each individual in the family tree.

1) Human pedigree for hair color

D = dark hair color; d = light hair color
In this case, light hair represents the "affliction".

Method of inheritance for the trait is: <u>simple Mendelian recessive.</u>

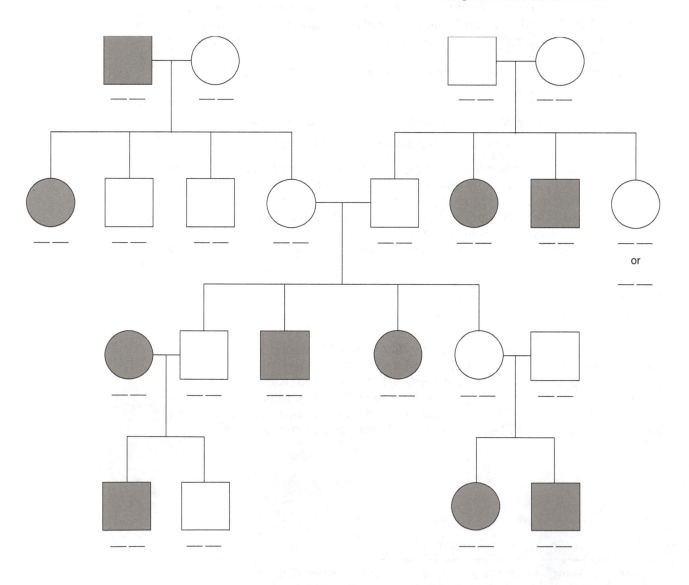

or

___ ___

2) Human pedigree for hair texture, an incompletely dominant trait
   H = straight hair; H' = curly hair
   Method of inheritance for the trait is: _____

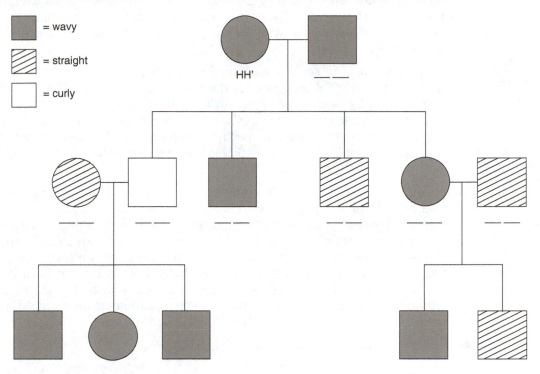

3) Human pedigree for muscular dystrophy, a sex-linked trait
   X^B = normal musculature; X^b = muscle atrophy
   Method of inheritance for the trait is: _____

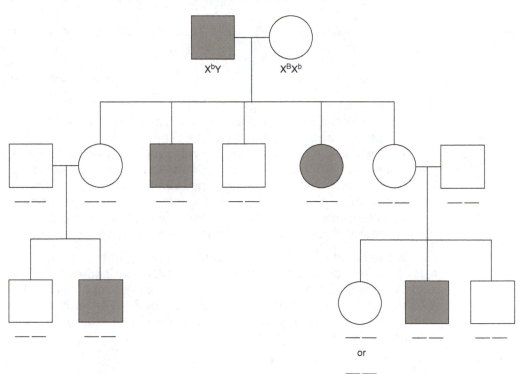

III. INTRODUCTION TO MOLECULAR GENETICS

Deoxyribonucleic acid (DNA) in the nucleus of the cell helps direct the synthesis of proteins. It does this by coding the exact sequence of **amino acids** which make up the proteins themselves. The code is **transcribed** in the nucleus and **translated** in cytoplasm where protein synthesis is completed. The process is described in more detail below.

A triplet code, or sequence of three **nucleotide** units on the DNA molecule, is used as (the original template for the transcription process. Each DNA code is transcribed by **messenger RNA** (mRNA) into a **codon.** This codon is also a sequence of three nucleotide units and is a direct complement to the original DNA triplet. Once an entire protein-specifying sequence of DNA nucleotides (a gene) is transcribed by mRNA, then the mRNA is carried to the sites of protein synthesis in the cytoplasm, the ribosomes.

Another form of RNA is also synthesized in the nucleus of the cell. This RNA is known as **transfer** RNA (tRNA). It is currently thought that there are over twenty different types of tRNA. Each tRNA molecule contains a group of three nucleotides called an **anticodon,** which is directly complementary to one mRNA codon. When the tRNA molecules pass from the nucleus to the cytoplasm they locate and bind to specific amino acids. It is the pairing of the tRNA anticodon with its complementary mRNA codon which causes the attached amino acids to be aligned in the proper sequence for protein synthesis. The amino acids are then enzymatically "glued" together in a binding process to form peptide bonds.

The process might look something like this:

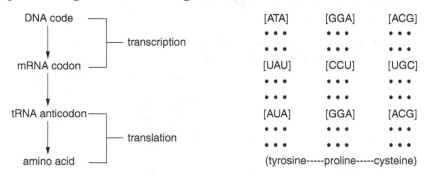

If you study the above example carefully, you will find definite patterns in the paired letters of the codons and anticodons. The letters are merely abbreviations of the **nitrogenous bases** attached to the nucleotides with A = adenine, T = thymine, G = guanine and C = cytosine. Adenine will always pair off with thymine and cytosine will always pair off with guanine. This particular complementary pairing of bases occurs only between the two strands which make up the DNA molecule. In the process of transcription into mRNA, a fifth base replaces thymine. This fifth base is called uracil (U). Uracil is not present in the DNA molecule itself, but is used in the synthesis of all forms of RNA during both transcription and translation.

For example, the DNA code ATA would be transcribed to the mRNA codon UAU; this codon would then be translated to the tRNA anticodon AUA. The tRNA molecule with the anticodon AUA would have already sought out and bonded with the specific amino acid (tyrosine) that it codes for. This amino acid would then be attached to the siring of amino acids previously formed in the process of protein synthesis.

DNA (Code)	mRNA (Codon)	tRNA (Anticodon)
(A) Adenine	(U) Uracil	(A) Adenine
(G) Guanine	(C) Cytosine	(G) Guanine
(T) Thymine	(A) Adenine	(U) Uracil
(C) Cytosine	(G) Guanine	(C) Cytosine

IV. ALPHAGENETICS

In this laboratory you will use a system called "alphagenetics" (an acronym for alphabet and genetics) to study the method by which a sequence of DNA codes are translated into proteins. Instead of using amino acids to build proteins, alphagenetics uses the letters of the alphabet to create words and phrases.

Procedure

In alphagenetics, instead of coding for an amino acid, each anticodon codes for a letter of the alphabet (see the list below). Just as in the genetic code, the alphagenetic code is repetitive. That is each letter of the alphabet is coded for by two to three anticodons.

Letter	tRNA Anticodons	Letter	tRNA Anticodons
A	AAA, GCU, CUA	N	GAG, UUC
B	AAG, GCC, CUG	O	GAU, UCA
C	AAU, UAA, CUU	P	GAC, UCG
D	AAC, UAG, CUC	Q	GGA, UCU
E	AGA, UAU, CCA	R	GGG, UCC
F	AGG, UAC, CCG	S	GGU, CAA
G	AGU, UGA, CCU	T	GGC, CAG
H	AGC, UGG, CCC	U	GUA, CAU
I	AUA, UGU	V	GUG, CAC
J	AUG, UGC	W	GUU, CGA
K	AUU, UUA	X	GUC, CGG
L	AUC, UUG	Y	GCA, CGU
M	GAA, UUU	Z	GCG, CGC

Example: Each anticodon codes for one of the twenty-six letters of the alphabet. By correctly transcribing (DNA to mRNA) and/or translating (mRNA to tRNA) the genetic code, you will obtain the proper alphagenetic letter. Each generic message can therefore be translated from the tRNA anticodons directly into a word or phrase.

For example, the sequence of DNA codes below is translated into the word "help": the tRNA anticodon "AGC" codes for the letter "H", the tRNA anticodon "AGA" codes for the letter "E", and so on.

DNA:	AGC	AGA	ATC	GAC
mRNA:	UCG	UCU	UAG	CUG
tRNA:	AGC	AGA	AUC	GAC
	H	E	L	P

To test your comprehension of protein synthesis (specifically, the processes of transcription and translation), try to complete the alphagenetics problems presented below. Use the alphagenetic code provided to decipher the phrases hidden within the genetic codes presented to you in each problem. Write your answers in the spaces provided beneath each problem.

Note: Be sure to note the kind of strand that you are presented with prior to deciphering the code. Remember that thymine is replaced by uracil in both kinds of RNA.

Strand Type   **Strand Code**

1. tRNA: CCA UUG GUG AUA CAA   AUC UGU CAC AGA GGU
   letter: ___ ___ ___ ___ ___   ___ ___ ___ ___ ___

2. mRNA: CCA AGU   CCG GGG ACA CCA   ACA CCA
          AUU CUA AAC UAG AUA UCA UCU

   tRNA: ___ ___   ___ ___ ___ ___   ___ ___
          ___ ___ ___ ___ ___ ___ ___

   letter: ___ ___   ___ ___ ___ ___   ___ ___
           ___ ___ ___ ___ ___ ___ ___

3. DNA: GAA GTA GGC GCT AGT CCA TTC

   mRNA: ___ ___ ___ ___ ___ ___ ___

   tRNA: ___ ___ ___ ___ ___ ___ ___

   letter: ___ ___ ___ ___ ___ ___ ___

4. DNA: GGT AAT ATA AGA TTC CTT CCA   TGT CAA
        GAA CGT   TTG ATA CCG AGA

   mRNA: ___ ___ ___ ___ ___ ___ ___   ___ ___
         ___ ___   ___ ___ ___ ___

   tRNA: ___ ___ ___ ___ ___ ___ ___   ___ ___
         ___ ___   ___ ___ ___ ___

   letter: ___ ___ ___ ___ ___ ___ ___   ___ ___
           ___ ___   ___ ___ ___ ___

5. tRNA: CAG GAU GGC UCA GUU UAU GGG AGA
   GAG UCA CAG UGU GAG
   AUU GCU UUC GGU AAA CAA
   GCU GAG GCA UUU GAU UCC UAU

 letter: ___ ___ ___ ___ ___ ___ ___ ___
    ___ ___ ___ ___ ___
    ___ ___ ___ ___ ___ ___
    ___ ___ ___ ___ ___ ___ ___

6. DNA: CAG GGG AAA TTC GGT AAT TCC ATA TCG GGC TGT GAT TTC
 mRNA: ___ ___ ___ ___ ___ ___ ___ ___ ___ ___ ___ ___ ___
 tRNA: ___ ___ ___ ___ ___ ___ ___ ___ ___ ___ ___ ___ ___
 letter: ___ ___ ___ ___ ___ ___ ___ ___ ___ ___ ___ ___ ___

7. tRNA: UGA AGA GAG UAU GGC AUA AAU CAA
 letter: ___ ___ ___ ___ ___ ___ ___ ___

8. mRNA: CCG CCC UUU CUC CCA UAG CGA CCG UAU CUA CUC
 tRNA: ___ ___ ___ ___ ___ ___ ___ ___ ___ ___ ___
 letter: ___ ___ ___ ___ ___ ___ ___ ___ ___ ___ ___

9. Write out your full name in the spaces provided below. Then, using first column of tRNA anticodons provided in the alphagenetics list, encode your name through tRNA and mRNA to DNA.
 Name: ___ ___ ___ ___ ___ ___ ___ ___ ___ ___ ___ ___ ___
 tRNA: ___ ___ ___ ___ ___ ___ ___ ___ ___ ___ ___ ___ ___
 mRNA: ___ ___ ___ ___ ___ ___ ___ ___ ___ ___ ___ ___ ___
 DNA: ___ ___ ___ ___ ___ ___ ___ ___ ___ ___ ___ ___ ___

Tissues and Skin

I. INTRODUCTION

A tissue is a group of cells operating together to perform a specific function. The cells making up a tissue are generally all derived from the same embryological precursor and are usually found in close proximity to one another. The study of tissues is called **histology.** There are some 200 different kinds of cells in the human body, but fear not. You don't have to learn them all. At least in this course you don't. These cells are grouped into only four main kinds of tissue:

epithelial
connective
muscle
nerve

In this exercise you will examine the structure of epithelial, connective, muscle tissues, and nerve tissue.

For each of the types of tissue listed below, obtain a prepared slide and examine it under low and high power. In conjunction with your examination, consult the photographs in the texts available in the laboratory, as well as the diagrams in this manual. As you study each tissue, make note of the most characteristic features, and make a sketch of several cells. Remember there may be more than one cell type and tissue type per slide.

II. EPITHELIAL TISSUES

Epithelial tissue **covers body surfaces, lines nearly all body cavities** and **forms glands.** The cells of epithelial tissue are close together and are joined by various types of junctions, such as desmosomes, tight junctions and gap junctions.

A. Simple Squamous Epithelial Tissue

This tissue consists of a single layer of flat, scale-like cells. It lines the **air sacs** or **alveoli** of the lungs, **glomerular capsule of the kidneys,** crystalline lens of the eye, and the eardrum. Simple squamous tissue that lines the heart, blood vessels and lymphatic vessels and forms **capillary** walls is called **endothelium.** When it lines the body cavity and covers viscera it is called **mesothelium.**

The slide you are looking at may be a bit of simple squamous epithelium placed flat on the slide. If this is the case, you will see regularly shaped cells in face view, all joined together. Or the slide may be similar to the one you made in another course when you scraped the lining of your mouth and examined the cheek cells.

From *Human Biology Laboratory Exercises* by Bray et al. Copyright © 1997 by Kendall/Hunt Publishing Company. Used with permission.

If this is the case, the cells will be dissociated from one another, and they will appear irregular in shape. Some of the cells will be folded over on top of themselves. In any event, sketch a small bit of simple squamous epithelium in the appropriate space at the end of this exercise, and make a few notes about its appearance.

B. Simple Cuboidal Epithelial Tissue

This tissue consists of a single layer of **cube-shaped cells.** When sectioned at right angles, the cuboidal nature of the cells is obvious. It covers the surface of the **ovaries**, lines the inner surface of the cornea and lens of the eye, forms part of the **tubules in the kidneys**, lines the smaller ducts of some **glands**, and forms the secreting units of other glands.

The slide you will study was either made from the kidney or from a gland. Look for small circles, and locate the cuboidal cells which form the circle. On some slides, some of the circles seem to be lined with columnar cells. See the description below. Sketch a bit of simple cuboidal epithelium on the page provided and make notes of its most characteristic features.

C. Simple Columnar Epithelial Tissue

This tissue consists of a single layer of **columnar cells.** When sectioned at right angles, these cells appear as **rectangles**. This tissue is adapted for **secretion and absorption**. It lines the **stomach, small and large intestines, digestive glands,** and **gall bladder.** In these locations, the cells protect underlying tissues. Some of the cells are modified in that the plasma membranes are folded into **microvilli.** This associated structure increases the surface area for absorption of food in the small intestines. Other cells are modified as **goblet cells** that store and secrete **mucus** to protect the lining of the gastrointestinal tract.

The slide you will study is most likely a cross section of some part of the digestive tract. This is a good time to remember how these slides are made. Pieces of animal tissue are cut out, processed, and cut into very thin slices. It is fairly easy to make cross sections of the digestive tract. It is impossible to prepare a slide containing only simple columna epithelium. To do this one would have to dissect away all the other layers of the gut, leaving only the single layer of cells which line the gut. So there are other tissues on your slide besides the one you are being told to study at the moment. One of the reasons there is an instructor in your lab is to help you find the part of the slide you need to study. If the instructions aren't enough help, ask for help!

On your gut slide, the epithelial tissue is the lining of the lumen of the gut. Examine the whole slide under 4X and locate the inside, or lumen. Use 10X and 40X to examine the single layer of cells which lines the lumen. The goblet cells are named for their shape, and tend to look blue because the mucus they contain stains blue with the dyes usually used. You may be able to find a goblet cell in the act of spitting out its glob of mucus. Sketch a bit of this epithelium, including a goblet cell, and make notes on its appearance.

D. Pseudostratified Epithelial Tissue

This tissue **appears to be stratified** since nuclei are viewed at various levels within the layer rather than at the base of each cell as in simple columnar epithelium. It lines much of the **upper respiratory tract, oviducts** and **uterus** and has **cilia** on its surface. Mucus formed by the associated **goblet** cells forms a thin film over the surface of the tissue, and the **movements of the cilia propel substances** over the surface of the tissue. The associated structure of cilia look like a thin area of ragged fringe on the luminal surface of the cells.

The slide you will study is most likely a cross section of trachea. As with the intestine, there are many other kinds of tissue on this slide. Be sure to get help if you cannot find the epithelial cells. Sketch a bit of tissue, and make notes.

E. Stratified Squamous Epithelial Tissue

This tissue consists of **several layers of cells**. The **superficial or top cells are flat**, whereas the **deep cells vary** in shape from **cuboidal to columnar**. The basal cells continually multiply by cell division. As surface layers are sloughed off, new cells replace them from the basal layer. The surface cells of keratinized stratified squamous epithelium contain a waterproofing protein called **keratin.** Keratin also resists friction and bacterial invasion. It should come as no surprise that the outer layer of the skin is keratinized. The surface cells of **nonkeratinized** stratified squamous epithelium do not contain keratin. The nonkeratinized variety lines wet surfaces such as the mouth, esophagus, and vagina. You studied it when you looked at the cheek lining cells. Sketch a bit of stratified squamous epithelial tissue, making sure to show the layers of cells.

F. Transitional Epithelial Tissue

This tissue consists of **several layers of cells**. When relaxed these cells appear to be cubodial but when they are stretched they will appear to be squamous cells. These particular epithelial cells are **specialized to stretch.** They are found in the **urinary bladder** and they allow this organ to store urine for disposal at a later time. The slide you will be using will be from the urinary bladder or possibly part of the ureters. Try to find an area that is relaxed and one that is stretched to the fullest. Sketch a bit of this tissue.

III. CONNECTIVE TISSUE

Connective tissue is the most abundant tissue in the body and functions in **protection, storage, support** and **binding of structures together**. It is different from epithelial tissue in that the cells are usually not closely joined together, but are separated from one another by relatively large amounts of extracellular material. This extracellular material may be hard like the matrix of bone, fibrous like the collagen fibers of tendons, or liquid like blood plasma.

A. Loose (Areolar) Connective Tissue

This is one of the most widely distributed connective tissues in the body. It consists of a viscous extracellular substance, three kinds of fibers, and several kinds of cells. The cells in loose connective tissue include fibroblasts and macrophages. Macrophages and some other cells of the loose connective tissue are part of the immune system and help guard against infection. Next time there is a chicken in your kitchen, pull up a bit of skin and notice the thin web which holds the skin in place. This is loose connective tissue.

B. Adipose Connective Tissue

This is fat tissue in which the fibroblasts are modified for **fat storage**. These cells are called **adipocytes.** Because the cytoplasm and nuclei of the cells are pushed to the side, the cells resemble signet rings. The fat is removed during the processing of the tissue for sectioning, but the hole remains. The tissue provides **insulation**, a **food reserve, support,** and **protection.** Sketch a bit of adipose tissue, and label the space left by the fat globule.

**Epithelial tissue
cell types:**

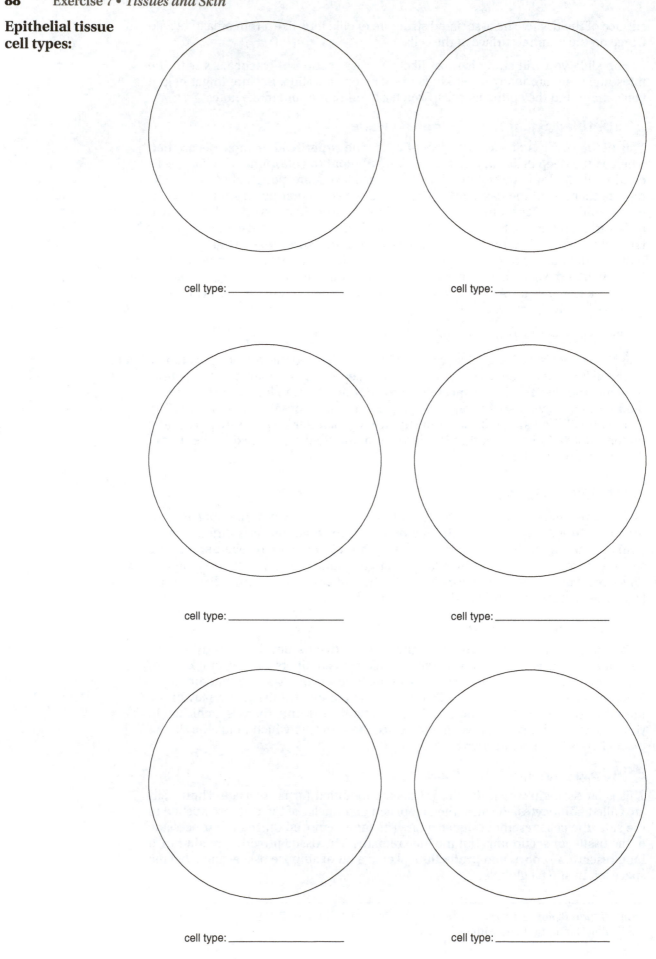

cell type: _____

cell type: _____

cell type: _____

cell type: _____

cell type: _____

cell type: _____

C. Hyaline Cartilage

All types of cartilage consist of a dense network of collagenous and elastic fibers firmly embedded in a gel-like substance. The cells of cartilage, called **chondrocytes,** occur singly or in groups in spaces called **lacunae** in the extracellular substance. Hyaline cartilage is the most abundant cartilage in the body. It provides **strength and support**. It is found at **joints** over the ends of the long bones, and the ventral ends of the ribs, and helps stiffen the **nose, larynx, trachea,** and **bronchi**.

The slide you will study may well be the old familiar cross section of trachea you used for pseudostratified epithelium. The cartilage is a sort of rainbow colored tissue with purple globs in it. If you have a bit of trachea, look for adipose tissue also, just for review. Sketch a bit of cartilage, and make notes on its appearance.

D. Bone

In bone, the extracellular material is about 70% hard inorganic salts, mostly **calcium phosphate**, and about 30% fibrous proteins, mostly **collagen**. The cells are in little pockets, called **lacunae,** just as they were in cartilage. It is important to remember that the cells made the extracellular matrix, cementing themselves permanently into little cells in the process.

Two kinds of bone slide may be available, decalcified and ground. The decalcified bone has been treated to remove the calcium phosphate matrix, and thus shows the cells better. The ground bone was cut from a cleaned dead bone, and does not show the cells at all, but the matrix is very clear. If possible you need to study both slides and construct an imaginary picture of how a real bone appears in life. Begin with the decalcified bone and start with 4X. It is a longitudinal section through a long bone of a mouse, and should look quite a bit like figure A in the bone diagram at the end of the book. Notice the **spongy bone** which fills the end near the joint. You may be able to see the **epiphyseal plate,** which marks the area of bone growth in young animals. The large cavity in the middle part of the bone is the **marrow cavity.** Red Bone Marrow is located in the spongy bone and produces all blood cells not just red blood cells. The marrow cavity of the long bone is full of fat. Blood vessels from the marrow cavity penetrate the bone to supply the blood cells with food and oxygen, and to collect their garbage. These blood vessels run in small canals, called **Haversian canals.** One of the common problems students have in studying bone is to confuse the marrow cavity with Haversian canals. A long bone has only one marrow cavity, and it is macroscopic. (This means BIG.) A bone has many tiny Haversian canals, which can be seen only with the microscope. Sketch the low power view of the decalcified bone, and make notes on its appearance.

Look at the part of the bone marked with the box in the diagram, and go up to high power. Here you can see the little purple Haversian canals, with bone cells arranged around them in concentric rings. How many rings of bone cells surround one Haversian canal? The blood vessels in the Haversian canals supply the **OSTEOCYTE** or bone cells with food and oxygen, and collect their garbage. So no bone cell can be too far away from a Haversian canal. Sketch the high power view of the decalcified bone and make notes on its appearance.

Now look at the slide of ground bone. It should look very much like figure C on the diagram. Here the cells are not visible, but the extracellular matrix is clearly seen. The **lacunae** which contained cells in life, are now empty. You can also see the tiny canals, called **canaliculi,** which allow nutrients and wastes to diffuse through the matrix from one cell to another. Sketch the appearance of ground bone, and make notes.

We tend to think of bone as an old dead tissue which never changes once we stop growing. This is very far from the truth. Bone is alive. The cells are alive. Bone has blood vessels and nerves, and bleeds when broken or cut, and hurts, as anyone who has ever had a broken bone can testify. Even in adults bone is continually being synthesized by **osteoblasts** and broken down by **osteoclast** bone cells.

E. Blood

Since blood is composed of cells separated from one another by a large amount of extracellular material, it is considered to be a connective tissue. In this case the extracellular material happens to be the liquid, **plasma.**

Examine a blood smear and locate the red and white blood cells. The red blood cells, **erythrocytes,** appear pink. They have no nuclei and are biconcave in shape. Erythrocytes are the most numerous of all blood cells and constitute about 40% of whole blood (cells and plasma).

White blood cells, **leukocytes,** are more scarce and generally larger than erythrocytes. They are divided into two groups: **granulocytes,** which have protein granules within their cytoplasm that are colored with various stains; and **agranulocytes** which have fewer granules. In a Wright's stain the nuclei will be stained purple.

There are three types of granulocytes. **Neutrophils** are the most numerous leukocytes, have a lobed nucleus with 2 to 5 parts, and granules which stain light pink. **Eosinophils** are hard to find as they usually represent only 1 to 3% of all circulating leukocytes. They have a dark-staining bilobed nucleus and deep red-staining granules. **Basophils** are even more scarce than eosinophils, and have a bilobed nucleus, and dark blue-staining granules.

The two types of agranulocytes are **lymphocytes** and **monocytes.** Lymphocytes constitute 25 to 33% of all leukocytes and are only slightly larger than erythrocytes. They have a large round dark-staining nucleus that nearly fills the cell, so there is only a thin circle of cytoplasm present. **Monocytes** are the largest blood cells and are two to three times the size of erythrocytes. Their large nuclei assume numerous shapes.

Search the slide for these cells. Sketch and label each cell with its distinguishing characteristics.

Connective tissue cell types:

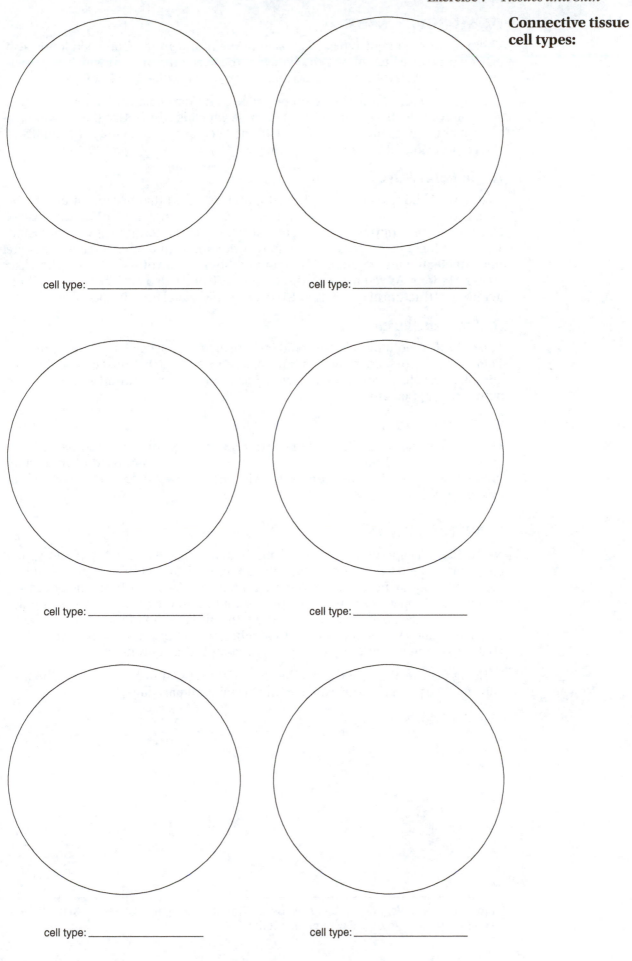

cell type: _____

cell type: _____

cell type: _____

cell type: _____

cell type: _____

cell type: _____

IV. MUSCLE TISSUE

There are three major types of muscle tissue found in the body; **skeletal** (also called striated) which moves the bones of the skeleton, **smooth** which is associated with the walls of blood vessels and organs, and **cardiac** or heart muscle.

All three types of muscle are on one slide in no particular order. First examine all three types with 4X and 10X to determine which is which. Use the diagrams in the appendix to help. Use the instructor as necessary. Then study the details of each type, using the information below.

A. Skeletal Muscle

First a word about vocabulary. Skeletal muscle is called that because it moves the skeleton. But it also has cross striations and is often called **striated** muscle. However, cardiac muscle also has cross striations so the terminology is confusing. The individual muscle cells are very long with a uniform diameter. These cells are **multinucleate** and are often called muscle **fibers** or myotubes. Notice the clear cross **striations.** As you recall, these are the result of overlap of thick and thin fibers in the ultrastructure of the muscle. Sketch a bit of skeletal muscle and make notes.

B. Smooth Muscle

The cells of smooth muscle are spindle-shaped, and each cell has its own nucleus. There are no cross striations. Smooth muscle makes up the muscular wall of the gut, bladder, uterus and blood vessels. Sketch a bit of smooth muscle and make notes on its appearance.

C. Cardiac Muscle

This is the muscle of the heart. It has **striations**, but its structure is somewhat different from that of skeletal muscle. Cardiac muscle is composed of individual uninucleate cells which are connected with **intercalated discs**. Make a sketch of this tissue.

V. NERVE TISSUE

Nerve Tissue is specialized to carry impulses. It is found in the brain and spinal cord that make up the **Central Nervous System** (CNS) and in the ganglia, spinal nerves and cranial nerves of the **Peripheral Nervous System** (PNS). The **neuron** is the cell that is specialized to carry impulses. It has **dendrites** to carry the impulse into the nerve cell body and **axons** to carry the impulse out of the nerve cell body. The axon may pass the impulse on to another neuron or onto a muscle fiber. The **glial** cells of the nervous tissue support and protect the neurons.

Using your textbook and any other material that you may find draw a neuron and label the parts. Give an explanation of what each part does.

Neuron:

**Muscle tissue
cell types:**

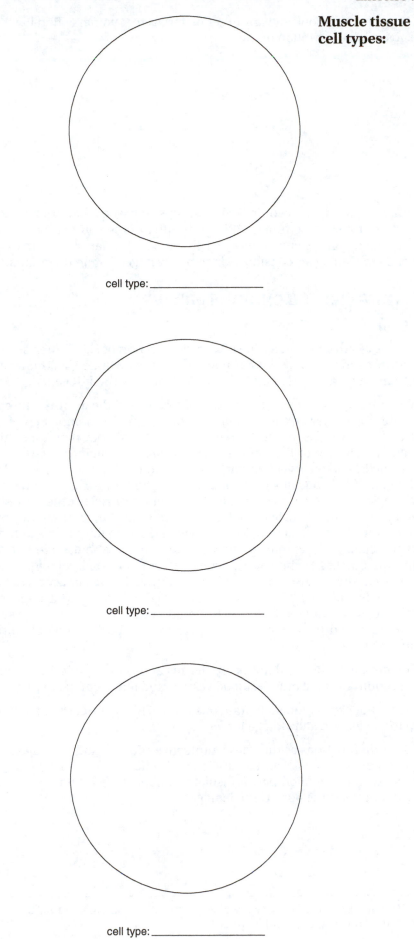

cell type:_____

cell type:_____

cell type:_____

Using your textbook and any other materials that you may find list the glial cells and give a job description of each.

Organs—An organ is composed of groups of tissues that are working together to carry out a specific function. In subsequent labs you will be studying many organs and organ systems. Today we will examine human skin as an example of an organ. Many biologists consider skin to be the largest organ that humans have.

VI. SKIN—THE CUTANEOUS MEMBRANE

A. Skin

Focus the microscope on a slide of a vertical section of skin. When sections of skin are cut for slides it is likely that many structures will be sliced unevenly. Under these circumstances only parts of various structures can be seen.

Skin is composed of two layers, the **epidermis** and the **dermis.** The **epidermis** is the outermost layer and it is composed of stratified squamous epithelium. Note that the epidermis is much thinner than the dermis but that at certain places it invaginates deep into the dermis. The two major invaginated structures are the **hair follicles** and the **sweat (sudoriferous) glands.** New cells are added by the root of the hair follicle pushing older ones toward the outside as hair grows. The elongated region of the hair is the **shaft.** Tiny smooth muscles called **arrector pili** are attached to the hair follicle in such a way that when they shorten they pull the hair into a more upright position. This activity is used mainly to make the fur into a thicker insulating layer in cold temperatures but it is also used to fluff the fur as a display that makes the individual appear larger than usual. An outpocketing of the hair follicle that is also composed of epidermis is the **sebaceous gland.** These glands secrete oil that lubricates the hair into the follicle. These appear as clumps of cells adjacent to the follicle. They should not be mistaken for the sweat glands which are coiled tubules and will appear as clumps of cuboidal epithelial cells arranged in circles.

The dermis is thicker than the epidermis and is deep to it. In this layer can be seen sections of some muscle, blood vessels, and many connective tissue types.

Below the skin is the subcutaneous layer. This layer contains adipose that is used for insulation and storage for fat.

Your skin functions as the body's protective coating, and although it is not hard or exceptionally tough, it forms an extremely effective barrier against penetration by disease-causing microbes. It insulates and cushions the underlying body tissues and protects the body from damage.

1. Using a magnifying glass, examine the back side of your hand. The epidermis is folded here to allow the skin to stretch. Look for **epidermal grooves,** diamond-shaped areas. They are especially noticeable around your knuckles.
2. Look at your fingertips with the magnifying glass. Finger prints are **epidermal ridges**, an area of the epidermis where the skin is folded to allow for traction. Your toes have these same ridges. Press your thumb into the ink pad provided and on the lab report sheet roll your inked thumb to produce a thumb print. Do the same with your middle finger. Compare your prints with those of your lab partners. There are three basic fingerprint patterns, whorls, loops, and arches. Which is most common in your prints?

Fingerprints are inherited. Individual family members will have similar, but not identical finger prints.

B. Hair

A hair is a column of dead epidermal cells covered with a layer of flattened, scale-like cells, the cuticle. Beneath the cuticle of the hair is the **cortex** and **medulla.** In this area you will find pigments (usually melanin) and air spaces. The air spaces increase with age, changing not only the color of the hair but also its texture. Because of the way the air spaces scatter light, they cause hair to gray as you age.

The part of the hair that shows above the skin's surface is the hair **shaft.** The rest of the hair, the root, is embedded in the skin, in the **hair follicle.** Humans have three distinct kinds of hair: fetal hair, continuously growing hair, and hair that grows to a certain length and then stops.

1. Obtain a prepared microscope slide labeled adult scalp. Under low power find a hair follicle with a hair in it. At the bottom of the follicle is a bulb shaped structure. It is in the bulb that new hair growth occurs. Look at the cells in the bulb under high power. Now follow up the hair shaft and see if you can identify the cuticle, medulla, and cortex. Sketch a hair shaft in #3 on the postlab questions.

At any one time, about 10% of the hair follicles on your scalp are in a resting stage where no new growth is occurring. These are the hairs that fall out when you brush or comb your hair. The structure of the hair depends upon the configuration and internal diameter of the hair follicle. Hair is curly if the follicle is bent, and straight if the follicle is straight. Coarse hair has a shaft diameter that is considerably greater than that of fine hair.

C. Nails

Plates of hard keratin (a fibrous protein) cover the dorsal surface of the end of the fingers and toes. These are your nails. Each nail has three parts. The **free edge** juts out over the tip of the finger or toe and is the part we file or clip. The **root** of the nail is buried beneath the skin at the base of the nail. Beneath the nail is the **nail bed.** The nail grows as cells are formed in the nail bed.

1. Observe an unpolished nail with a magnifying glass or under a dissecting microscope. Notice that the nail is made of layers of keratin. Nails are pink because of the numerous blood vessels underlying the nail. The white half moon area, most noticeable on your thumbs, is an area of particularly thick epidermal cells.

From *Human Biology: A Laboratory Manual*, Fifth Edition/Revised Printing by Roberta B. Williams. Copyright © 2000 by Kendall/Hunt Publishing Company. Used with permission.

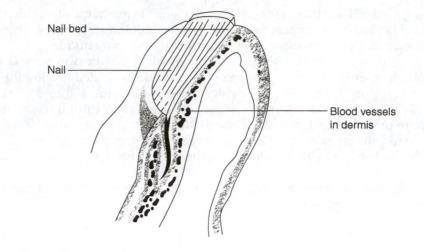

Nail bed

Nail

Blood vessels
in dermis

Figure 7.1 Cross section through a nail.

1. Sketch what your skin looks like under the magnifying glass at the knuckles
   and finger tips.

2. Put a fingerprint in the space below. Do you have whorls, loops or arches in
   your fingerprint?

3. Sketch a hair shaft. Label the cuticle, medulla and cortex of the hair shaft.

4. Sketch the surface of a fingernail as seen under the magnifying glass.

5. Explain how structure and function were related in the things that you viewed in this lab exercise.

Locate these anatomical terms on the figure provided.

1. epidermis
2. dermis
3. subcutaneous layer
4. adipocytes
5. hair bulb
6. sweat/sudoriferous gland
7. sebaceous/oil gland
8. muscle
9. blood vessels
10. nerve
11. pores
12. stratified squamous cells

Figure 7.2 Integument structure. © 2003 Mark Nielsen. Art by Jamey Garbett.

VII. WHY WE SHOULD USE SUNSCREEN

Objective

In this exercise, you will be studying the effects of Ultraviolet light exposure on bacterial growth.

Materials

- 21 Petri plates with nutrient agar
- ultraviolet lamp boxes
- timers
  cultures of *Bacillus megaterium, Serratia marcescens,* and *Staphylococcus aureus*
- disposable inoculation loops or swabs

Procedure

1. The instructor will assign each student a specific organism and exposure time. Record your information below:

   Organism: _____ Exposure time: _____

2. Label the bottom (agar in it) of the plate—your initials, the organism and exposure time.

3. Streak the entire surface of the nutrient agar, using a swab. See diagram below.

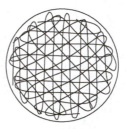

4. Insert the Petri plates into the ultraviolet lamp boxes, with the covers removed. Using an index card cover one half of each plate and expose the uncovered side for the time specified by your instructor.

   Note: DO NOT LOOK DIRECTLY AT THE ULTRAVIOLET LIGHTS.

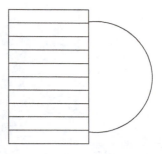

5. After your exposure is completed, recover your plate, wrap in aluminum foil, and incubate agar side up at 25° C for 48 hours.
6. When you return to lab class next week, record your results. Compare the unexposed bacteria to the UV exposed bacteria. Compute the survival rate by estimating the amount of bacteria left on the UV exposed portion of the petri plate.

   Enter your results below your exposure time.
   Full coverage— +++
   can see some agar— ++
   can see more agar than bacteria— +
   no bacteria— 0

Organism	*Exposure Times*						
Bacillus megaterium (endospore former for protection)	30 sec	1 min	2 min	5 min	10 min	15 min	30 min
Serratia marcescens (a red bacteria found in soil)	10 sec	20 sec	30 sec	1 min	2.5 min	5 min	10 min
Staphylococcus aureus (a white bacteria found on us)	10 sec	20 sec	30 sec	1 min	2.5 min	5 min	10 min

Using your textbook and any other material that you may find draw each type of epithelial tissue cell listed. Give the location and/or organs where it is found in the human body, characteristics or jobs that the cell performs and if applicable associated cells and structures such as goblet cell, cilid and microvilli.

A. 1. SIMPLE SQUAMOUS   DRAWING

    2. (location)

    3. (characteristic)

B. 1. SIMPLE CUBODIAL

    2. (location)

    3. (characteristic)

C. 1. SIMPLE COLUMNAR

    2. (location)

    3. (characteristic)

    4. (associated structure)

    5. (associated cell)

D. 1. PSEUDOSTRATIFIED COLUMNAR

    2. (location)

    3. (characteristic)

    4. (associated structure)

    5. (associated cell)

E. 1. STRATIFIED SQUAMOUS

    2. (location)

    3. (characteristic)

F. 1. TRANSITIONAL

    2. (location)

    3. (characteristic)

Using your textbook and any other material that you may find draw each type of connective tissue listed. Give the location and/or organs where it is found in the human body, characteristics or jobs that the cell performs and if applicable associated cells and structures.

A. 1. ADIPOSE   DRAWING

    2. (location)

    3. (characteristic)

    4. (cell name)

B. 1. BLOOD

    2. (location)

    3. (characteristic)

    4. (cell names)

       a.

       b.

       c.

C. 1. BONE

    2. (location)

    3. (characteristic)

    4. (cell name)

D. 1. CARTILAGE

    2. (location)

    3. (characteristic)

    4. (cell name)

    5. (associated types—give the locations)

       5A.   HYALINE

       5B.   FIBROCARTILAGE

       5C.   ELASTIC

Using your textbook and any other material you may find, draw each type of muscle fiber (cell) listed. Give the location and/or organs where it is found in the human body, characteristics or jobs that the cell performs and if applicable associated cell structures such as strations, nuclei and interalloted disks.

DRAWING

A. 1. Cardiac

2. (Location)

3. (Characteristics)

4. (Associated cell structures)

B. 1. Skeletal

2. (Location)

3. (Characteristics)

4. (Associated cell structures)

C. 1. Smooth

2. (Location)

3. (Characteristics)

4. (Associated cell structures)

Skeletal System

I. INTRODUCTION

In this exercise you will learn to identify the bones of the human skeleton. In some cases you will also learn some of the bone extensions (**processes** or **condyles**) and openings (**foramina**). These are important as places for attachment of muscles and tendons and for passage of nerves, blood vessels, and ligaments through bone respectively. They will also serve as landmarks that will help you distinguish one bone from another.

The human skeleton consists of **206** bones. It is divided into the **axial skeleton** which consists of the skull, vertebrae, ribs and sternum; and the **appendicular skeleton** which includes the bones of the appendages and the girdles which support them.

AXIAL SKELETON

II. SKULL

The skull consists of the **cranium** which houses and protects the brain and the facial region.

A. Cranium

The **frontal bone** forms the forehead and includes part of the roof of the nasal cavity and orbits. Articulating posteriorly with the frontal bone are the paired **parietal bones.** These form much of the top and sides of the head. The **occipital bone** articulates with the posterior of the parietals and forms the rear and part of the base of the cranium. The **foramen magnum** is a large hole in the occipital bone through which nerves pass from the brain and become the spinal cord. On either side of this foramen are the **occipital condyles** which articulate with the first vertebra. Forming the side and part of the base of the skull in the ear region are the paired **temporal bones.** These articulate with the parietals above and the occipital posteriorly. The opening in this bone is the **external acoustic meatus** through which sound waves travel to the middle and ultimately the inner ear. Although the inner ear itself cannot be seen, the **otic capsule** which houses it can be viewed on the inside surface of the temporal bone. The **internal acoustic meatus** can be seen as an opening in the otic capsule where the cranial nerve takes the sound information from the inner ear to the brain. Below the external acoustic meatus are the **mastoid** and **styloid processes** which serve as attachment points for some muscles. Extending anteriorly from the external acoustic meatus is the **zygomatic process** which articulates with the **malar (zygomatic) bone** to form the zygomatic arch. This is a point of attachment for some of the muscles used for chewing. Just between the zygomatic process and the external acoustic meatus is a depression called the **mandibular fossa** where the lower jaw articulates with the skull. On the bottom of the skull and articulating anteriorly with the occipital bone is the **sphenoid bone.** This forms much of the floor of the cranium and extensions on either side form part of the orbit. When viewed from above with the top of the skull removed the shape of the sphenoid is often compared to a bat with outstretched wings. Also in this view, the **sella turcica** can be seen. This is an area where the bone rises up forming a saddle-like depression. The sella turcica is where the pituitary gland rests during life. The **ethmoid bone** is anterior to the sphenoid but it is difficult to view. The **perpendicular plate** and **superior nasal conchae** of the ethmoid can be seen in an anterior view of the nasal cavity. A very small part of the ethmoid can be seen on the nasal surface of each orbit. The **cribriform plate** of the ethmoid bone can be viewed from above. The tiny holes in this plate are the **olfactory foramina** which allow nerves for the sense of smell to pass from the nasal cavity to the olfactory lobes of the brain. An extension of the ethmoid called the **crista galli** separates the right and left olfactory lobes.

B. Facial Region

The paired **maxillae** form the upper jaw and contain sockets for the upper teeth. They have processes that extend medially to form much of the hard palate which is the roof of the mouth and the floor of the nasal cavity depending on your perspective. The **palatine bones** form the posterior of the hard palate and extend upward to form part of the rear walls of the nasal cavity. The malar bones which were mentioned before are considered to be facial bones and are located between the temporal bone and the maxilla on the zygomatic arch. The **lacrimal bone** is a small bone on the nasal side of each orbit. There is a groove in each through which a lacrimal duct transports lacrimal fluid (tears) from the eye to the nasal cavity. The **nasal bones** form the bridge of the nose. The **vomer** forms part of the nasal septum and a very small section of it is visible on the bottom of the skull just posterior to the hard palate. The **inferior nasal conchae** are the lowermost conchae in the anterior view into the nasal cavity. The **mandible** is the lower jaw and the articulation of the **mandibular condyle** with the mandibular fossa of the temporal bone is the only movable joint between two skull bones. The mandibular condyle should not be mistaken for the **coronoid process** which is the site of attachment of some jaw muscles.

III. HYOID BONE AND AUDITORY OSSICLES

The **hyoid** bone is located in the ventral neck region and does not articulate with any other bones. Consequently, it is most easily identified on mounted skeletons where it is held in place by wire. It is attached to the larynx and parts of the skull by muscles and ligaments and its major role is to move the larynx during swallowing.

There are three pairs or six auditory **ossicles**: the **malleus** (hammer), **incus** (anvil), and the **stapes** (stirrup). These are found in the middle ear and are important in hearing. The ossicles are the smallest bones in the human body.

IV. VERTEBRAL COLUMN

The human vertebral column consists of **33 vertebrae** that in life surround and protect the spinal cord among other functions. Most vertebrae have several characteristics in common. Most have a **body** that is ventral to the spinal cord and a **spinous process** that is dorsal. Typical vertebrae also have paired **transverse processes.** They will have paired **anterior articulating surfaces** that connect to paired **posterior articulating surfaces** of the vertebra in front. As adjacent vertebrae articulate they are cushioned by the **intervertebral discs.** The articulation of vertebrae also forms the **intervertebral foramina** that allows spinal nerves to pass to and from the spinal cord.

The 33 vertebrae are divided into smaller groups according to distinctive characteristics and region. The most cranial seven are called **cervical vertebrae.** They are all characterized by paired **transverse foramina** through which arteries pass to the brain. No other vertebrae have them. The first two cervical vertebrae are distinctive. The **atlas** articulates cranially with the occipital condyles of the skull. Its joint allows up and down head movements like nodding. The atlas articulates posteriorly with the axis and this joint allows side to side head movements. The next twelve are the **thoracic vertebrae.** On the sides of these are the **rib facets** where the ribs articulate. The next five vertebrae are **lumbar.** Perhaps the best way to distinguish these is to eliminate the transverse foramina and the rib facets and look for a very heavy body. The **sacrum** is a fusion of the next five vertebrae. It articulates with the pelvic girdle and this fusion allows this joint to support more weight. The four **caudal vertebrae** of the human fuse into a small tail bone called the **coccyx.**

V. RIBS AND STERNUM

The human has twelve pairs of **ribs.** The upper seven articulate ventrally with the various parts of the **sternum** by way of their **costal cartilages,** and dorsally with the thoracic vertebrae. The five most caudal rib pairs are called false ribs because they do not attach directly to the sternum. Rib pairs 8, 9, and 10 share attachment to the sternum with the costal cartilage of rib 7. Rib pairs 11 and 12 are not associated with the sternum and are called floating ribs.

There are three parts to the sternum. The cranial part is the **manubrium** which articulates laterally with the **clavicle** and first two rib pairs. The middle sternum is the **body** and the lowermost part is the **xiphoid process.**

VI. APPENDICULAR SKELETON

A. Pectoral Girdle

The pectoral girdle is composed of a **clavicle** and **scapula** on each side. The clavicles articulate medially with the manubrium and laterally with the **acromion process** of the scapula. The scapula is a triangular bone with a **spine** running across its dorsal surface. The spine ends laterally at the acromion process. Near the acromion process and extending from the anterior border is the **coracoid process.** Both processes are sites of attachment for arm and chest muscles. In a lateral view the **glenoid cavity** which articulates with the humerus can be seen.

B. The Upper Appendage

The **humerus** is the most proximal bone of the upper limb and its rounded **head** fits into the glenoid cavity. At its distal end the rounded **capitulum** articulates with the radius of the forearm, and the grooved **trochlea** articulates with the **semilunar (trochlear)** notch of the **ulna.** The proximal end of the ulna is the **olecranon** or elbow to which are attached muscles that extend the forearm. During this action the elbow fits into the **olecranon fossa** of the humerus. At the distal end of the ulna is a small **ulnar styloid process** that is a bump that can be felt on the little finger side of the arm just proximal to the wrist. The proximal radius articulates with the capitulum of the humerus and the **radial notch** on the lateral aspect of the ulna near the semilunar notch. At its distal end a small **radial styloid process** can be felt on the thumb side of the forearm. When the hand is pronated (palm down) the radius is lateral to the proximal end of the ulna, but its distal end crosses the ulna to end in a medial position. When the hand is supinated (palm up) the head of the radius rotates on the capitulum turning the entire bone to a position lateral to the ulna.

The hand is composed of 8 wrist bones called **carpals,** 5 palm bones called **metacarpals,** and 14 finger bones called **phalanges.**

C. The Pelvic Girdle

The pelvic girdle is composed of 2 **innominate bones** that articulate posteriorly with the sacrum and with each other anteriorly at the **pubic symphysis**. Each innominate bone is composed of three bones that fuse together before birth. The uppermost portion is the flared **ilium** that forms the hip. This area articulates with the sacrum. The anterior portion of the innominate is the **pubis** where articulation between the innominates occurs. The lowermost part is the thick **ischium.** Between the ischium and pubis is the large **obturator foramen** where some nerves and tendons pass to the legs. On the lateral aspect of the innominate bone is the cup-like **acetabulum** where the head of the femur articulates.

D. The Lower Appendage

The **femur** is the largest bone in the body. At its proximal end is the **head** which articulates with the acetabulum. At its distal end are the **medial** and **lateral condyles** that articulate with structures of the same name on the **tibia.** On the anterior surface of the distal femur is a groove called the **patellar surface** where the **patella** or kneecap fits. On the anterior surface of the proximal tibia is the **tibial tuberosity** where the patellar ligament attaches. The **anterior crest** extends most of the length of the tibia and serves as a place for muscle attachment. Medial to the crest is an area of tibia that is very close to the surface and is painful when scraped (shin). At the distal end of the tibia is the **medial malleolus** that can be felt just before the ankle begins. The **lateral malleolus** of the **fibula** can be felt in the same area only laterally. The two malleoli form an arch into which the ankle fits.

There are 7 ankle bones called **tarsals.** The **talus** articulates with the tibia and fibula. The largest tarsal is the **calcaneus** which forms the heel. The instep of the foot is composed of 5 **metatarsals** and there are 14 **phalanges.**

VII. SKELETAL TERMS

Bone Markings, Depressions & Cavities

1. Foramen: opening allowing passage of nerves or blood vessels
2. Fossa: shallow depression in a bone
3. Sulcus: groove or furrow
4. Meatus: canal or tube-like passage
5. Fissure: narrow slit
6. Sinus: cavity in a bone
7. Condyle: rounded, knuckle-like eminence on a bone that articulates with another bone
8. Tuberosity: large rounded process on a bone; serves as muscle anchor
9. Tubercle: small, round process
10. Trochanter: very large process
11. Head: process supported by a constricted part (neck)
12. Crest: narrow ridge of bone
13. Spine: sharp, slender process
14. Suture: line remaining after two bones have joined together and fused

15. Epicondyle: a small projection on or above a condyle
16. Facet: a smooth, flat articular surface
17. Fovea: a smooth depression; used most often for attachment or articulation
18. Lamina: a thin smooth plate of bone
19. Line: a low bony ridge for muscle attachment
20. Process: a bony projection
21. Ramus: a stem; an elongated process

VIII. REVIEW

Locate the following bones and structures on the articulated and disarticulated human skeletons. Then label the accompanying figures.

Axial Skeleton

frontal	posterior articulating surface	nasal
internal acoustic meatus	intervertebral discs	inferior nasal conchae
temporal	cervical vertebrae	mandibular condyle
occipital	atlas	hyoid bone
styloid process	thoracic vertebrae	spinous process
otic capsule	lumbar vertebrae	anterior articulating surface
malar (zygomatic)	occipital condyles	intervertebral foramina
sphenoid	parietals	transverse foramina
ethmoid	mastoid process	axis
superior nasal conchae	external acoustic meatus	rib facets
crista galli	foramen magnum	sacrum
maxilla	zygomatic process	coccyx
lacrimal	mandibular fossa	
vomer	sella turcica	ribs
mandible	perpendicular plate	manubrium
coronoid process	olfactory foramina	xiphoid process
body of vertebra	cribriform plate	costal cartilages
transverse process	palatine	body of sternum

Appendicular Skeleton

clavicle	ulna	talus
acromion process	olecranon fossa	phalanges
scapular spine	radial styloid process	pubic symphysis
humerus	metacarpals	ischium
capitulum		obturator foramen
radius	innominate bones	femur
olecranon process	ilium	patella
semilunar notch	pubis	medial and lateral condyles
carpals	acetabulum	of femur and tibia
phalanges	head of femur	fibula
scapula	patellar surface	anterior crest
coracoid process	tibia	lateral malleolus
glenoid cavity	tibial tuberosity	metatarsals
head of humerus	medial malleolus	calcaneus
trochlea	tarsals	

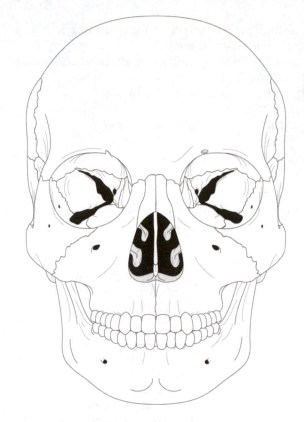

Figure 8.1   Cranium, anterior view. © 2003 Mark Nielsen. Art by Jamey Garbett.

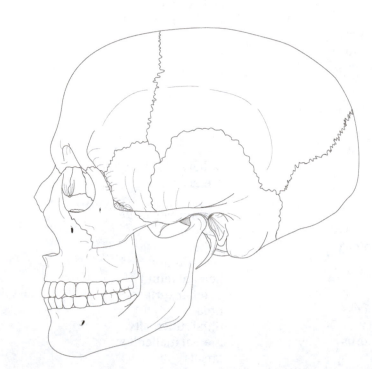

Figure 8.2   Cranium, lateral view. © 2003 Mark Nielsen. Art by Jamey Garbett.

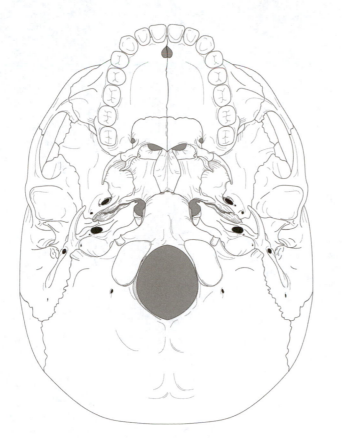

Figure 8.3   Cranium, inferior view. © 2003 Mark Nielsen. Art by Jamey Garbett.

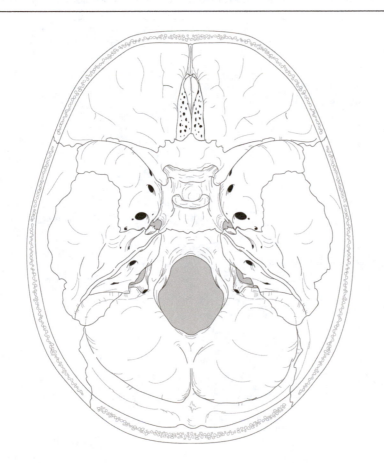

Figure 8.4   Cranial vault. © 2003 Mark Nielsen. Art by Jamey Garbett.

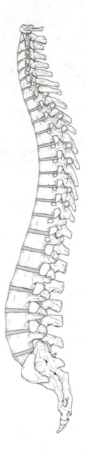

Figure 8.5   Vertebral column, lateral view. © 2003 Mark Nielsen. Art by Jamey Garbett.

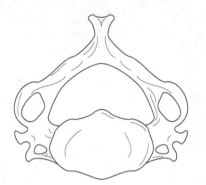

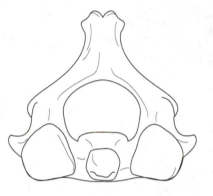

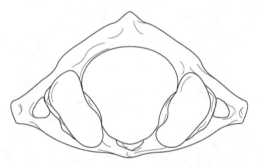

Figure 8.6   Cervical vertebrae, superior view. © 2003 Mark Nielsen. Art by Jamey Garbett.

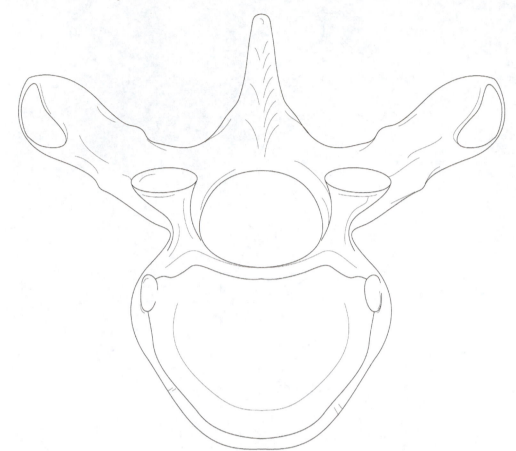

Figure 8.7   Thoracic vertebra, superior view. © 2003 Mark Nielsen. Art by Jamey Garbett.

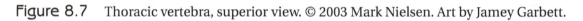

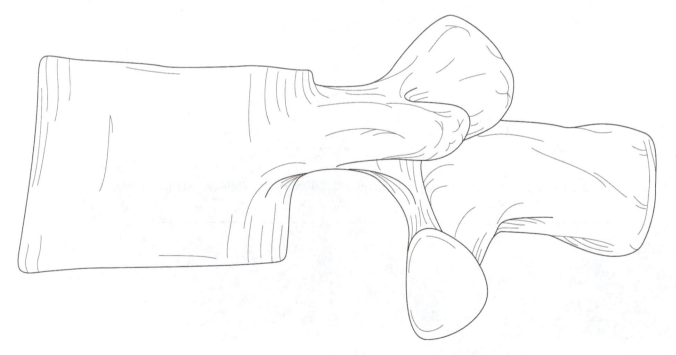

Figure 8.8   Lumbar vertebra, lateral view. © 2003 Mark Nielsen. Art by Jamey Garbett.

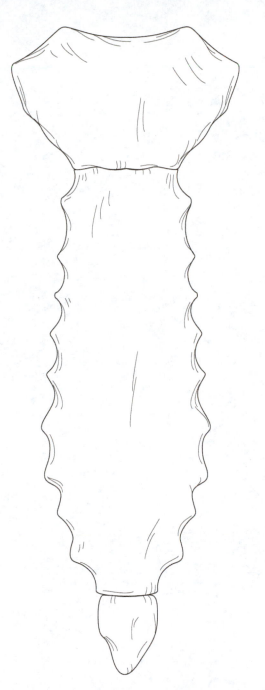

Figure 8.9   Sternum, anterior view. © 2003 Mark Nielsen. Art by Jamey Garbett.

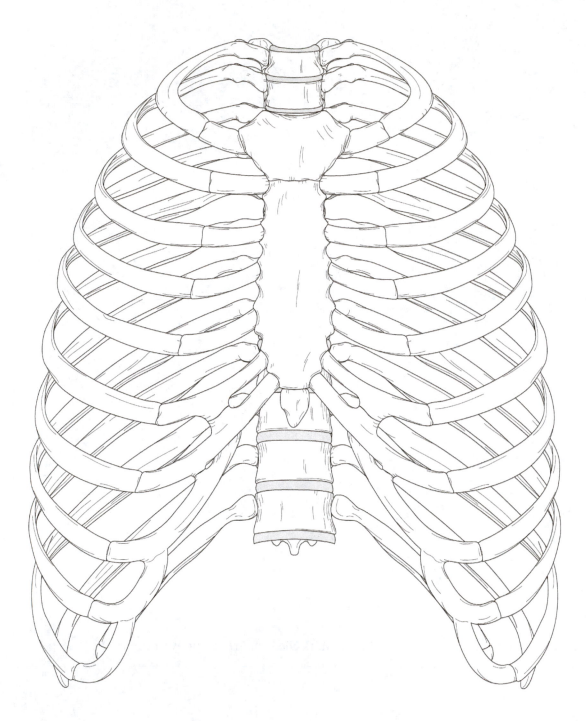

Figure 8.10   Thoracic cage, anterior view. © 2003 Mark Nielsen. Art by Jamey Garbett.

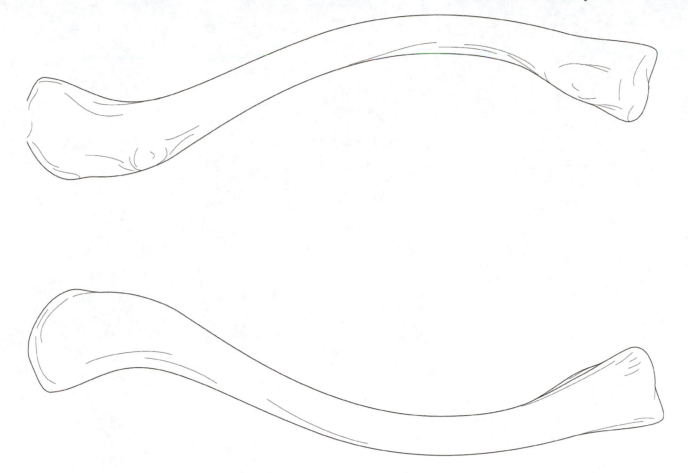

Figure 8.11   Clavicle. © 2003 Mark Nielsen. Art by Jamey Garbett.

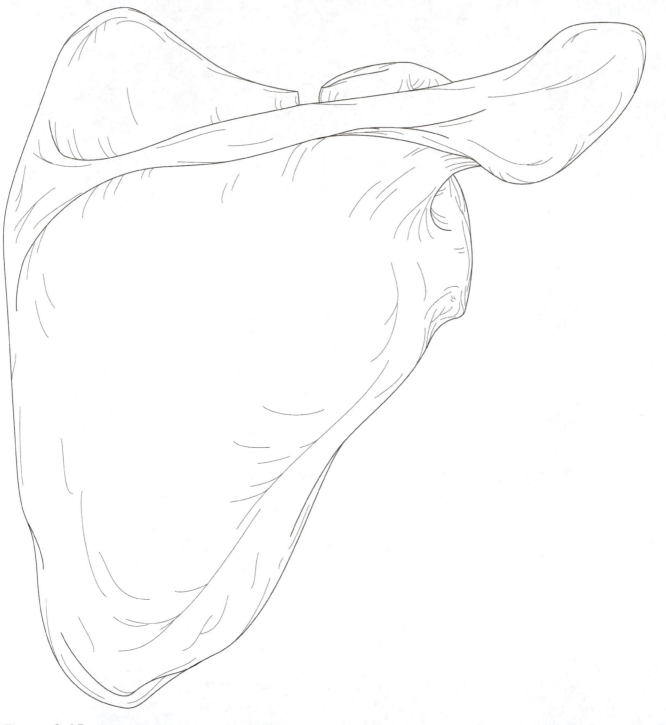

Figure 8.12   Scapula, posterior view. © 2003 Mark Nielsen. Art by Jamey Garbett.

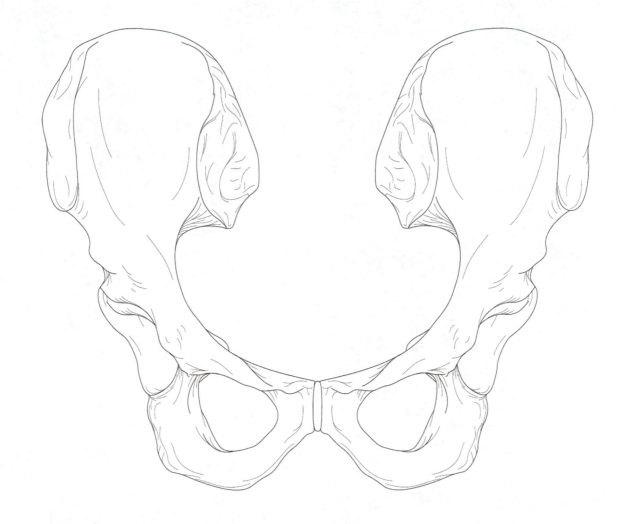

Figure 8.13   Os coxae articulated. © 2003 Mark Nielsen. Art by Jamey Garbett.

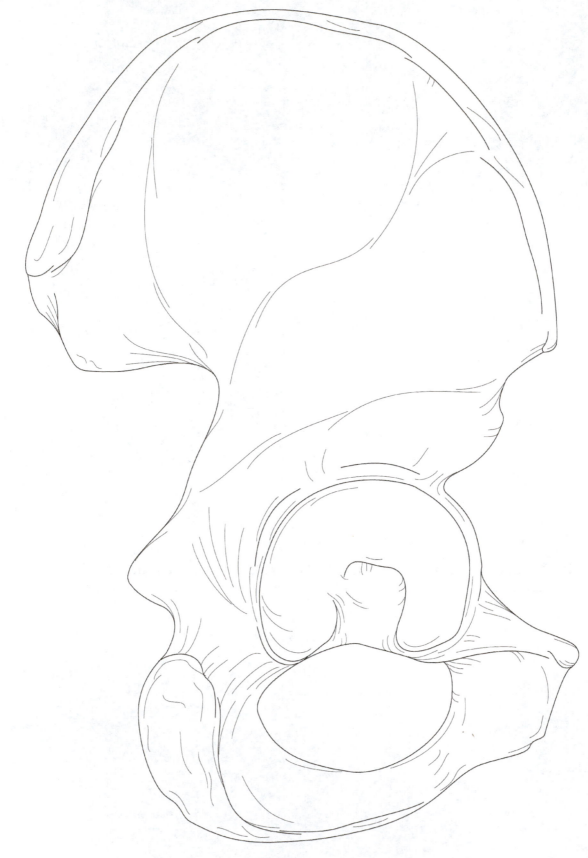

Figure 8.14   Os coxae, lateral view. © 2003 Mark Nielsen. Art by Jamey Garbett.

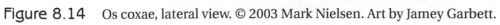

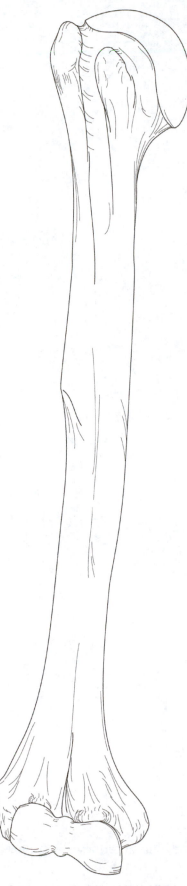

Figure 8.15   Humerus, anterior view. © 2003 Mark Nielsen. Art by Jamey Garbett.

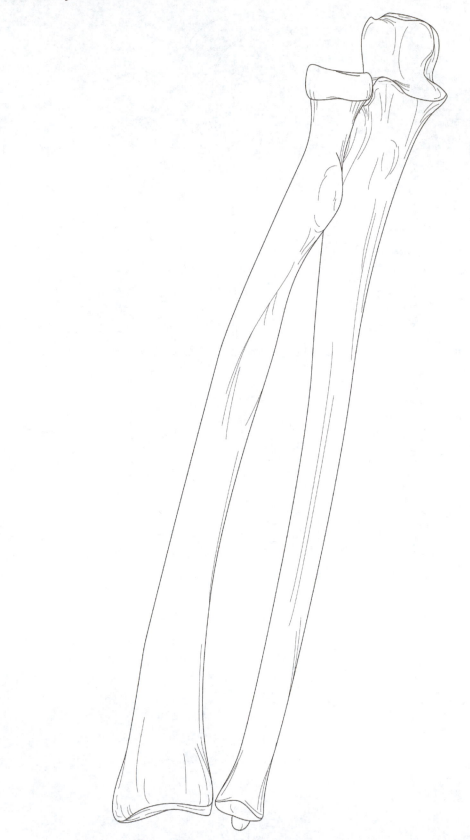

Figure 8.16   Radius and ulna, anterior view. © 2003 Mark Nielsen. Art by Jamey Garbett.

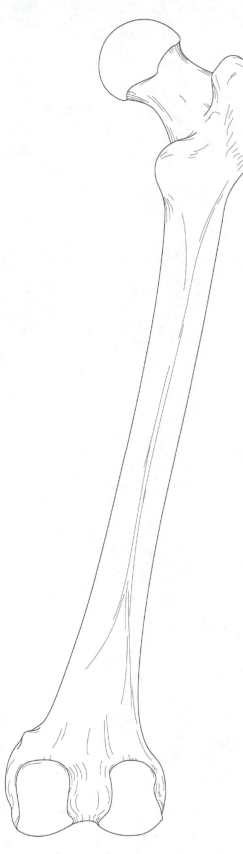

Figure 8.17   Femur, posterior view. © 2003 Mark Nielsen. Art by Jamey Garbett.

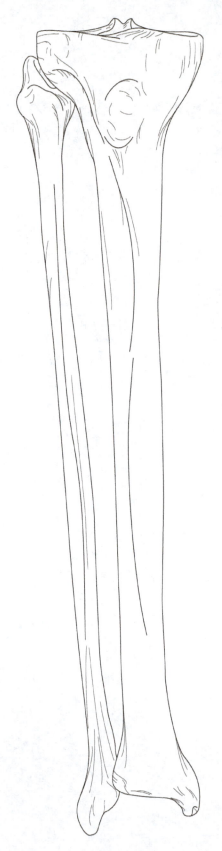

Figure 8.18   Tibia and fibula, anterior view. © 2003 Mark Nielsen. Art by Jamey Garbett.

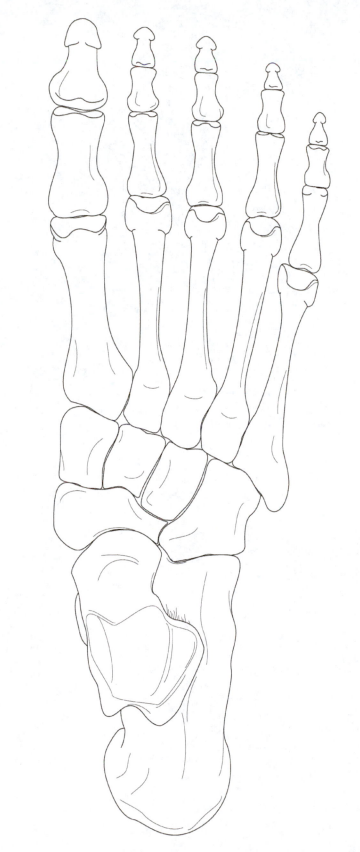

Figure 8.19   Foot, superior view. © 2003 Mark Nielsen. Art by Jamey Garbett.

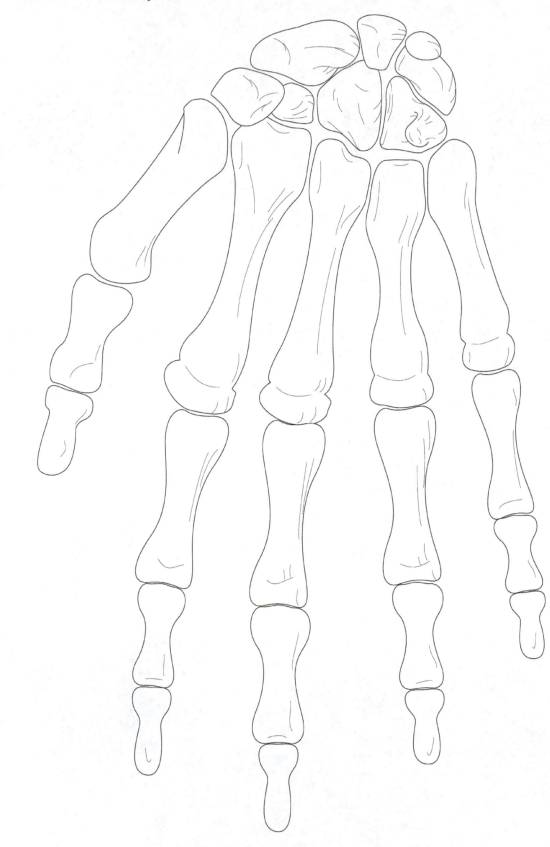

Figure 8.20   Hand, anterior view. © 2003 Mark Nielsen. Art by Jamey Garbett.

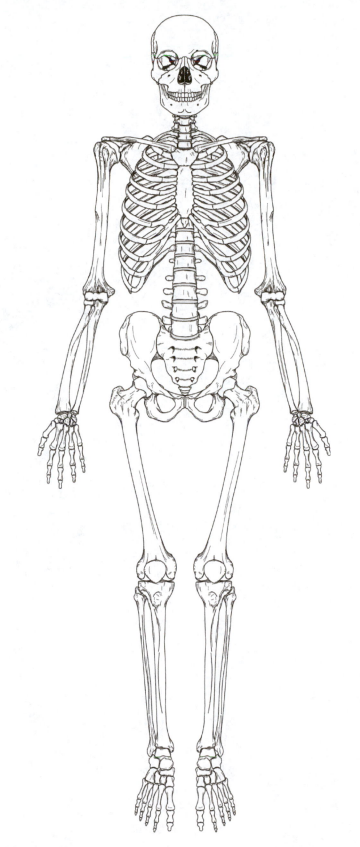

Figure 8.21   Skeleton, anterior view. © 2003 Mark Nielsen. Art by Jamey Garbett.

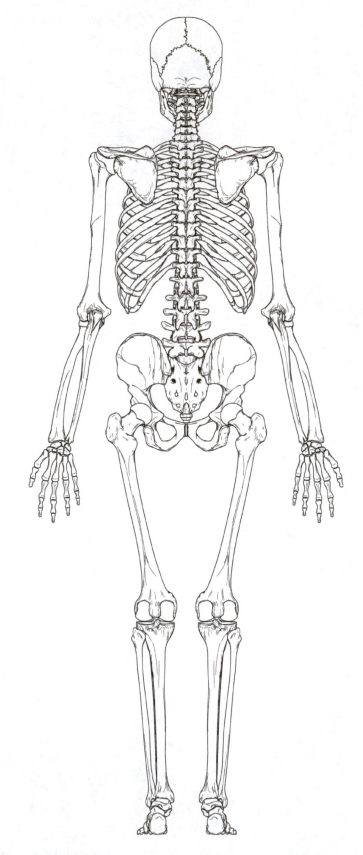

Figure 8.22   Skeleton, posterior view. © 2003 Mark Nielsen. Art by Jamey Garbett.

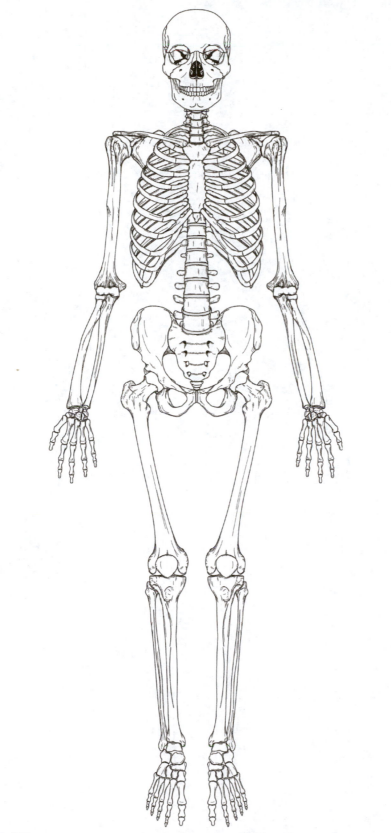

Figure 8.23   Skeleton, anterior view. © 2003 Mark Nielsen. Art by Jamey Garbett.

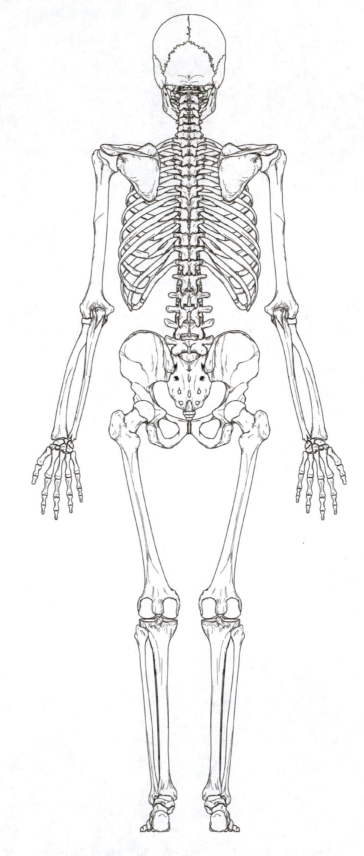

Figure 8.24   Skeleton, posterior view. © 2003 Mark Nielsen. Art by Jamey Garbett.

To test yourself, cut out the following bones and create an articulated skeleton.

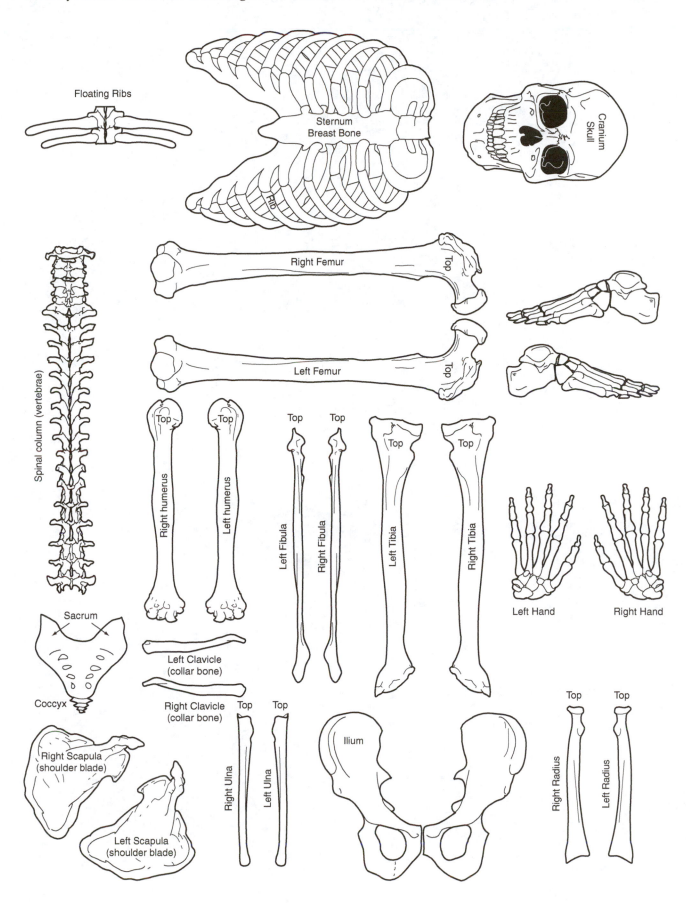

Floating Ribs

Sternum
Breast Bone

Rib

Cranium
Skull

Spinal column (vertebrae)

Right Femur

Top

Left Femur

Top

Right humerus

Left humerus

Left Fibula

Top

Right Fibula

Top

Left Tibia

Top

Right Tibia

Top

Left Hand

Right Hand

Sacrum

Coccyx

Left Clavicle
(collar bone)

Right Clavicle
(collar bone)

Right Scapula
(shoulder blade)

Left Scapula
(shoulder blade)

Right Ulna

Left Ulna

Top

Top

Ilium

Right Radius

Left Radius

Top

Top

Name: _____

Muscular System

I. INTRODUCTION

Skeletal muscles are attached to bones. Upon contraction, they will cause the bones to move. To create the constant movement and to handle the sometimes jarring of the bones, each muscle has to have a strong attachment at its origin and insertion. These attachments are at sites on the bones where there are rough surfaces, bulges, projections, etc. The connective tissue that attaches muscle to bone is called a **tendon**.

There are over **700** skeletal muscles of the human body. A complete study of the muscles involves not only the name and location of the muscles but also the action and the identification of their origin and insertion.

The **origin** of muscles is on a bony structure that is stationary. The muscle extends past at least one joint to attach to the moveable bone, the **insertion.**

Skeletal muscles only pull when they contract. They generally pull toward the point of origin. Skeletal muscles cannot push. Therefore, when the biceps brachii pulls causing the lower arm to flex, the biceps brachii cannot push the lower arm back to its original position. Another muscle must pull the lower arm back into position. That muscle is the triceps brachii. The biceps brachii and triceps brachii are **antagonistic muscles.**

Some muscles cause **flexion** which is a decrease in the angle between origin and insertion as when the arm is bent at the elbow. When the elbow is straightened the motion is **extension** and the angle between the humerus and the forearm increases. The foot can be flexed in two directions. If the toes are pointed out as when standing on tiptoe the angle between the calcaneus and the lower leg is decreased; this is **plantar flexion.** When the toes are flexed upward the angle between the metatarsals and the lower leg is decreased; this is **dorsiflexion.** If a body part is moved toward the midline of the body as in joining the hands in front of you the motion is **adduction.** Moving a part away from the midline as spreading the legs or raising the arms to the horizontal position is **abduction. Rotation** is the turning of a part around its axis. The arm can be kept at the side and rotated medially or laterally. If the arm is moved so that it forms a circular path the motion is called **circumduction. Elevation** is raising a body part and **depression** is lowering it. The jaw is elevated and depressed in chewing.

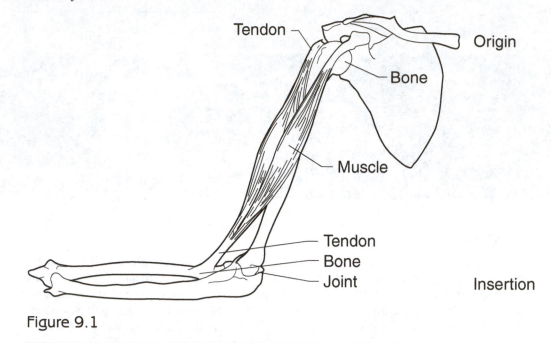

Figure 9.1

You should be able to define and perform the following actions.

flexion	extension	plantar flexion
dorsiflexion	adduction	abduction
rotation	circumduction	elevation
depression	pronation	supination

II. MUSCLE ASSIGNMENT #1

Using your text or information found online fill in the action and location for the following muscles:

Muscle	Action	Location
Buccinator		
Zygomaticus		
Orbicularis oculi		
Masseter		
Orbicularis oris		
Frontalis		
Occipitalis		

Muscle	Action	Location
Temporalis		
Sternocleidomastoid		
Rectus Abdominis		
Deltoid		
Teres major		
Latissimus dorsi		
Pectoralis major		
Trapezius		
Serratus anterior		
External oblique		
Biceps brachii		
Brachioradialis		
Flexor carpi radialis		
Triceps brachii		
Extensor carpi ulnaris		
Extensor carpi radialis		
Gluteus maximus		
Gluteus medius		
Sartorius		
Iliopsoas		
Adductor longus		
Rectus femoris		
Vastus medialis		

Muscle	Action	Location
Vastus lateralis		
Vastus intermedius		
Biceps femoris		
Semitendinosus		
Semimembranosus		
Gastrocnemius		
Peroneus longus		
Soleus		
Tibialis anterior		
Palmaris longus		

III. MUSCLE ASSIGNMENT #2

Using colored pencils, sketch the muscles on the diagrams provided. Indicate the origin and insertion.

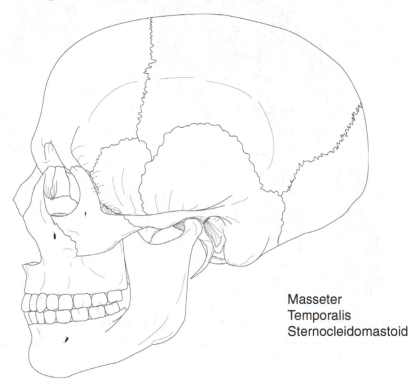

Masseter
Temporalis
Sternocleidomastoid

Figure 9.2 Cranium, lateral view. © 2003 Mark Nielsen. Art by Jamey Garbett.

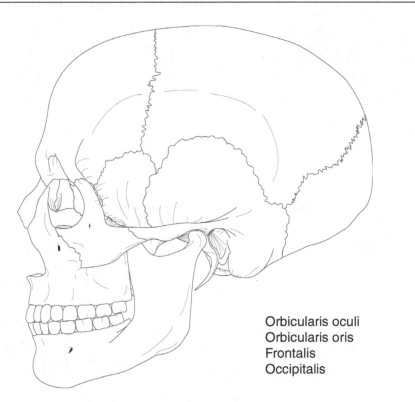

Orbicularis oculi
Orbicularis oris
Frontalis
Occipitalis

Figure 9.3 Cranium, lateral view. © 2003 Mark Nielsen. Art by Jamey Garbett.

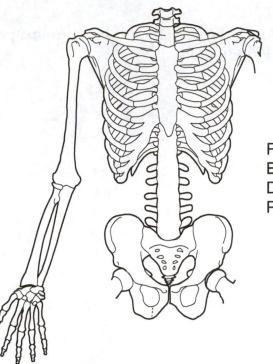

Rectus abdominus
External obliques
Deltoid
Pectoralis major

Figure 9.4 © 2003 Mark Nielsen. Art by Jamey Garbett.

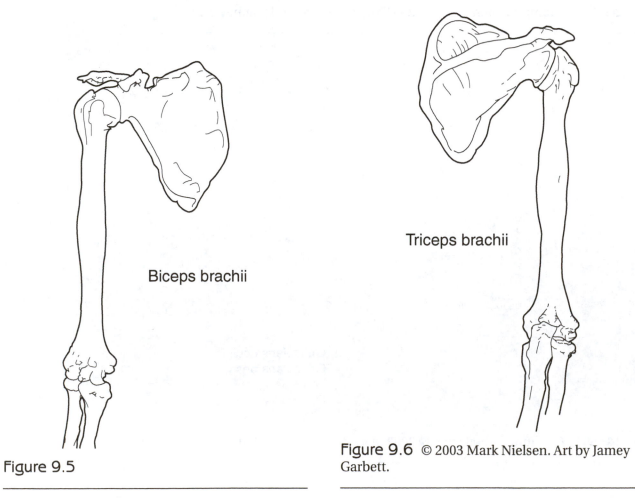

Biceps brachii

Triceps brachii

Figure 9.5

Figure 9.6 © 2003 Mark Nielsen. Art by Jamey Garbett.

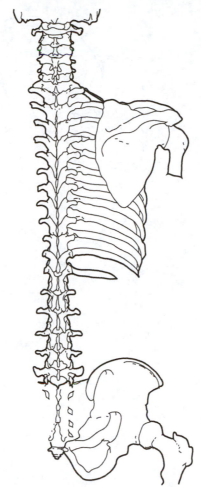

Trapezius
Latissimus dorsi
Deltoid
External obliques

Figure 9.7 © 2003 Mark Nielsen. Art by Jamey Garbett.

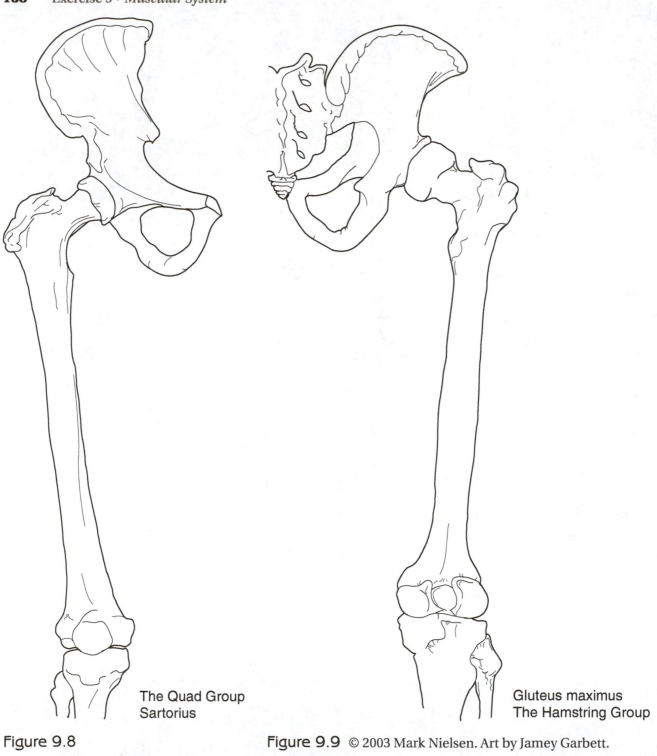

The Quad Group
Sartorius

Gluteus maximus
The Hamstring Group

Figure 9.8

Figure 9.9 © 2003 Mark Nielsen. Art by Jamey Garbett.

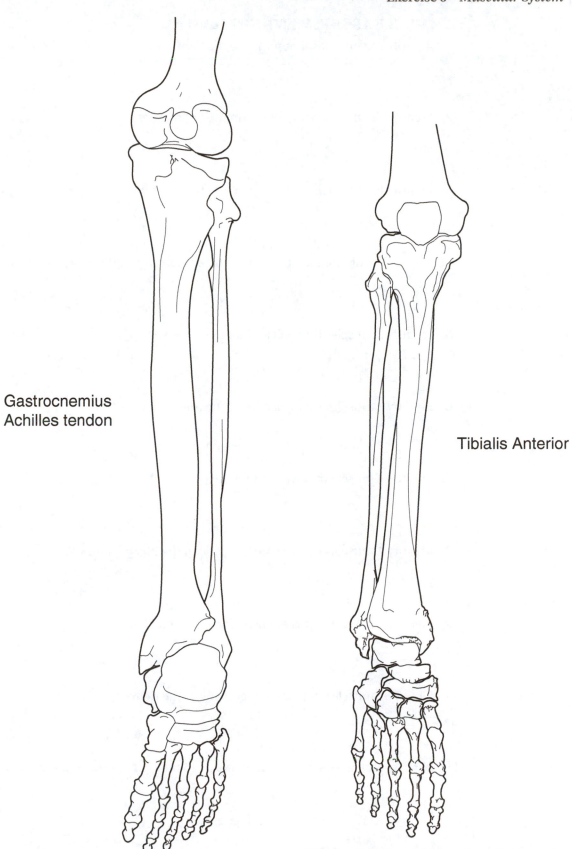

Gastrocnemius
Achilles tendon

Tibialis Anterior

Figure 9.10

Figure 9.11 © 2003 Mark Nielsen. Art by Jamey Garbett.

IV. MUSCLE ASSIGNMENT NUMBER #3

1. Identify the muscles that make up the hamstrings.

2. Identify the muscles that make up the quadriceps.

3. Identify the muscles that make up the rotator cuff.

4. What is the difference between the biceps femoris and the biceps brachii?

5. Identify 3 muscles that will adduct the humerus.

6. Identify 2 muscles that will adduct the thigh.

7. Identify a muscle that will abduct the thigh.

8. Identify 4 muscles that are antagonistic to the biceps femoris.

9. Identify a muscle that is antagonistic to the gracilis.

10. Identify 2 muscles that are antagonistic to the deltoid.

11. Identify one muscle that is antagonistic to the right sternocleidomastoid.

V. MUSCLE ASSIGNMENT #4

Locate these superficial muscles on figures 9.12 and 9.13.

Orbicularis oculi
Orbicularis oris
Frontalis
Occipitalis
Temporalis
Zygomaticus
Sternocleidomastoid
Rectus abdominis
Deltoid
Latissimus dorsi
Pectoralis major
Trapezius
External oblique
Biceps brachii
Triceps brachii
Brachioradialis

Extensor carpi radialis
Flexor carpi ulnaris
Gluteus maximus
Sartorius
Iliopsoas
Adductor longus
Rectus femoris
Vastus medialis
Vastus lateralis
Biceps femoris
Semitendinosus
Semimembranosus
Gastrocnemius
Gracillis
Soleus
Tibialis anterior

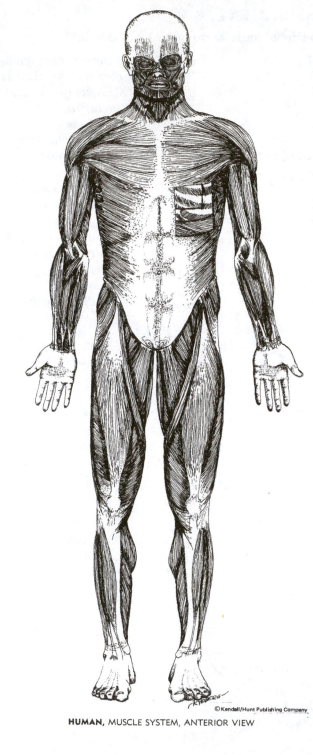

HUMAN, MUSCLE SYSTEM, ANTERIOR VIEW

Figure 9.12

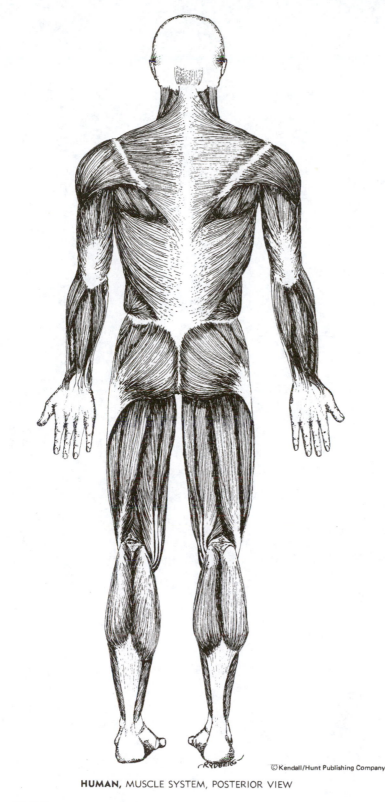

HUMAN, MUSCLE SYSTEM, POSTERIOR VIEW

Figure 9.13

Blood

I. INTRODUCTION

In addition to studying the structure and function of blood vessels and the heart, we will examine the fluid which the heart pumps and vessels carry, namely blood. **Hematology** is the study of blood. Blood is considered to be the most naturally nutritious substance known. It contains virtually everything needed to sustain life. In addition, blood is a medium for transportation.

Blood comprises about 6 to 8% of your body weight. The "average" female and male have about 5 quarts (4.6 *l*) and 6 quarts (5.6 *l*) of blood, respectively.

II. COMPOSITION OF BLOOD

Blood is composed of two portions: a liquid called **plasma** and **formed elements** which include various types of cells and platelets. Interestingly, blood is the only tissue composed of isolated (not connected) cells. More than 92% of plasma is water and about 7% is proteins. More than 70 different plasma proteins have been identified.

III. DETERMINATION OF HEMATOCRIT

One way to express the relationship between plasma and formed elements is the **hematocrit**. The hematocrit is the ratio of the volume of formed elements, mainly red blood cells, to the total blood volume.

What types of people would you expect to have higher than average hematocrits?

IV. OBSERVATION OF FORMED ELEMENTS IN THE BLOOD

Procedure

Obtain a prepared slide of human blood which has been treated with **Wright's stain.** First examine the slide under low magnification, then higher power. The most numerous cells are the round appearing **erythrocytes** (A.K.A. red blood corpuscles or RBC). Actually, RBC have a very precise, biconcave-disk shape, as if they were all punched out of the same mold. Exceptions to this shape and size are usually caused by medical problems such as sickle cell and giant cell anemias.

To state that the body "lives on oxygen" is scarcely an understatement. The major way oxygen is carried from the lungs to maintain life in each cell is by the hemoglobin contained in RBCs. Each RBC contains about 260 million molecules

---

of the blood pigment hemoglobin. Since RBC have the basic function of oxygen transport, it is not strange that when a person is deficient in hemoglobin, all organs suffer.

An average drop of blood contains about 250–275 million RBC. The average RBC count in females and males are 4,500,000 and 5,500,000 per mm^3 blood respectively.

People who possess RBC counts considerably less than these are said to be **anemic**. Anemia (literally means no blood) is caused by a variety of problems such as genetic (i.e., sickle cell), diet (vitamin B_{12}, folic acid, or iron-deficiency) and blood loss (hemorrhage). Carefully examine some RBCs under high magnification.

1. Do they possess a nucleus? _____
2. Sketch the shape of a RBC, seen from the top in panel A and in panel B viewed edgewise if cut in half.

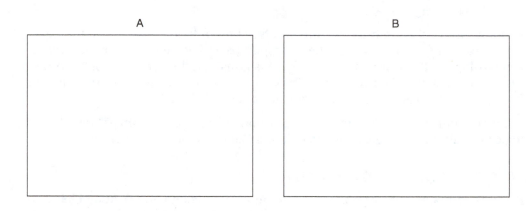

A   B

V. WHITE BLOOD CELLS—LEUCOCYTES

In addition to RBCs, your blood possesses **leucocytes** or white blood corpuscles (= WBC). Even though normal, healthy people have around 7,000 to 10,000 WBC per mm^3, they are difficult to find, since there are somewhere between 450 to 800 RBCs per one WBC! Under low magnification, WBC will appear nearly colorless. However, the Wright's stain allows the nucleus of the WBC to pick up color. With closer inspection under high power, you should be able to find five different types of WBCs.

Two of the five types have relatively large nuclei and do not possess granules in their cytoplasm. These **agranulocytes** are **lymphocytes** and **monocytes.**

Other WBCs have lobed nuclei and cytoplasmic granules, the **granulocytes**. We are able to distinguish three different granulocytes by how they appear following application of Wright's stain. **Neutrophils** (heterophils), **basophils** and especially **eosinophils** have rather beautiful coloration.

The most common WBC, neutrophils, have flexible membranes and are able to squeeze through capillary walls and like super-sleuths, seek out and engulf (phagocytize) pathogens. Eosinophil numbers increase during allergic reactions or parasitic infections. Lymphocytes, the second most abundant WBC, play an important role in immunity.

A major reason physicians take blood samples from an ill person is that a symptom of most infections is an **elevated** WBC count.

Lymphocyte	Monocyte	Neutrophil	Eosinophil	Basophil
20 to 25%	3 to 8%	60 to 70%	2 to 4%	0.1 to 0.2%
Spherical or slightly indented nucleus	Deeply indented or horseshoe-shaped nucleus	Lobated nucleus (Three to five lobes)	Lobated nucleus (Three to five oval lobes)	Elongated, lobated nucleus, often S-shaped
Thin peripheral layer of clear cytoplasm	Relatively large amount of clear cytoplasm	Uniformly fine cytoplasmic granules	Coarse spherical cytoplasmic granules	Coarse spherical cytoplasmic granules
Cytoplasm: Deep purple,	Cytoplasm: Muddy purple,	Cytoplasm: Lilac,	Cytoplasm: Lilac,	Cytoplasm: Lilac,
Nucleus: Pale blue	Nucleus: Pale blue	Nucleus: Pale lavender granules	Nucleus: Bright red granules	Nucleus: Deep blue granules

Figure 10.1 Characteristics of the five types of human leucocytes.

Procedure

Please attempt to locate at least 4 of the 5 types of leucocytes on your stained slide. Following confirmation of your identification from your instructor, place a check by the appropriate name.

neutrophil _____ eosinophil _____ basophil ____ lymphocyte _____ monocyte _____

Which leucocytes would be the easiest _____ and most difficult _____ to locate?

VI. PLATELETS AND CLOTTING

The circulatory system is unique in that it has self-repair capability. Circulating along with red and white blood cells are numerous small cell fragments called **platelets.** Platelets, like RBCs and WBCs, are produced by bone marrow. Platelets are not cells but rather appear as cell fragments which average about 250,000 per mm^3 of blood.

Platelets play a role in the formation of **clots.** Clots develop in response to various injuries which cause a blood vessel to be cut or torn. If these openings in the circulatory system are not quickly "sealed," a large blood loss called **hemorrhage** would occur.

In addition to platelets which release the chemical **thromboplastin** when broken, two other plasma proteins are required in the clotting process:

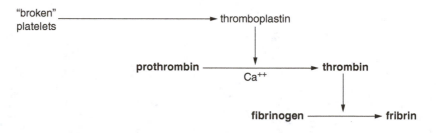

The fibrin strand or threads form the meshwork of the clot.

VII. DETERMINATION OF CLOTTING TIME

Procedures

1. Place a large drop of warm, engineered blood in the middle of a slide. The drop should be about 5 mm in diameter. Place the slide on the light box.
2. With a toothpick draw a line through the blood drop at 30-second intervals until a fibrin strand is seen to adhere to the toothpick. Avoid the edge of the drop. Normally it will take from 3 to 6 minutes for the blood to clot.

Results_____ minutes.
People who possess blood which does not quickly clot could have an inherited, i.e. genetic, disease called?_____

REVIEW QUESTIONS

Please answer the following questions.

1. _____ What is the predominant cell in blood?

2. _____ Give the scientific name for WBCs.

3. _____ Which cell functions in clotting?

4. _____ What is an average hematocrit value?

5. _____ Which cell functions in oxygen transport?

6. _____ Name the chemical which prevents blood from clotting.

7. _____ Which is the most abundant WBC?

8. _____ Describe each WBC functioning in immunity.

    A) Granolocytes: B) Agranolocytes:

      a) neutrophils   a) monocytes/monophages

      b) basophils   b) lymphocytes

      c) eosinophils   i) B cells

                                          ii) T cells

9. _____ Name the condition where RBC count is low.

10. _____ What is the name of common blood stain?

11. Given the following:
    a. 1000 mm^3 = 1 ml; b. average person has 5,000,000 RBCs per mm^3; and c. average person has 5 liters (= 5000 ml) blood; calculate approximately how many RBCs an average person has in her/his circulating system.

      Answer _____ RBCs.

12. Would you expect people who live at higher elevations to have higher RBC counts? Explain.

13. Do you think that Olympic athletes (or any others) should be allowed to increase their RBC count by (a) training at altitude or (b) the process of "blood doping"?

14. Who is Dracula? What is his association with hematology?

Cardiovascular Assessment

I. INTRODUCTION

If for no other reason, the cardiovascular system warrants study because of its role in human health. About 700,000 people in the U.S. die each year of heart attacks (**myocardial infarction**). This is the number one killer of Americans! Fatalities caused by heart attacks and failure of blood vessels (including strokes) together cause more than 50% of all deaths. This is three times more deaths than those caused by cancer each year in the U.S. Probably every adult alive knew someone who died owing to a cardiovascular problem. Many of these deaths can be attributed to "bad" lifestyle such as smoking, poor diet, obesity, and lack of exercise. Eliminating these aspects from your life is a way that you can take an active role in promoting good health. Warning—bad habits and cardiovascular problems usually begin in the late teens and early twenties. Unfortunately most people do not take such warning "to heart" until it is too late!

In this laboratory, we will conduct several important clinical assessments in hopes that you will understand how these **diagnostic** aids work. Remember your laboratory instructor is not a medical doctor so she/he cannot officially diagnose any clinical problems. We will listen to heart sounds, record electrocardiograms, measure heart rate and blood pressure and will compare our measured values to **population norms.**

II. HEART RATE MEASUREMENT

Each time your ventricles contract and force blood into your arterial "tree," your arteries first expand and then recoil. This expansion and recoil of the arterial tree can be felt externally as a **pulse.** Consequently, a determination of the pulse rate is an indirect measurement of the heart rate. Pulse rate can be determined at any site where a large artery can be pressed against a bone or firm tissue. The most common sites are the wrist and neck.

Using one of your forefingers, carefully take your own pulse (at either the wrist or neck) by counting the number of beats during one minute. Record this number in the space provided on your laboratory Review Questions sheet. Then perform the other indicated calculations. The average **resting** heart rate for a healthy young adult is about 72 beats per minute. How does your heart rate compare? _____

Physicians can easily diagnose three conditions from measuring pulse rate: **arrhythmia** which simply means that the time between pulses is not equal, **bradycardia** which is a slow heart rate, and **tachycardia** which is a fast heart rate. Bradycardia is characterized by a resting heart rate of less than 50 beats per minute while tachycardia is characterized by a resting heart rate greater than 100 beats per minute. Do you fall into any of the three categories? _____

III. HEART SOUNDS

During the cardiac cycle, closing of the heart valves makes the familiar and hopefully distinct "**lubb-dubb**" sound. The initial sound "lubb" is produced by the closing of the A-V valves. Often additional sounds are apparent which usually are called heart murmurs. Heart **murmurs** appear as extra squeaking, "shhhhing", or rumbling noises. Many are caused by diseased or faulty valves which allow blood to move back and forth instead of traveling smoothly through the heart. Some murmurs result simply from increased turbulence as the blood flows through the heart and are not medical problems.

The heart is located near the middle of the thoracic cavity, not on the left side as usually seen during the Pledge of Allegiance. These people have their hand over their left lung, not their heart! Given this information, discretely use a **stethoscope** to listen to your heart. Each "lubb-dubb" equals one heart beat. Do you hear any heart sounds in addition to the expected "lubb-dubb"? _____

IV. BLOOD PRESSURE

To overcome the resistance of the vessels particularly the arterioles and the capillaries to blood flow, blood must be forcefully ejected from the heart. The greater the resistance in the vessels, the harder the heart must pump to push blood through the vessels. Each time that the left ventricle contracts forcing blood into the systemic aorta, blood is at its highest pressure. This is termed the **systolic pressure.** When the ventricle relaxes, pressure in the arterial tree falls. The lowest pressure during this cycle is called the **diastolic pressure.** Blood pressure is reported as systolic over diastolic, such as 125/75. The units of pressure are always in millimeters of mercury (mm Hg).

Physicians measure blood pressure indirectly with a **sphygmomanometer.** The upper arm is the common site for this determination since the **brachial artery** is readily available and near to the heart.

Procedure

1. The pressure of the blood itself can be indirectly, accurately, and noninvasively measured by the use of a special apparatus known as a **sphygmomanometer.** Obtain a sphygmomanometer and a stethoscope, and prepare to take your laboratory partner's blood pressure. When you have completed the procedure indicated below, exchange positions with your laboratory partner and repeat the procedure. Use electronic sphygmomanometers if they are available.
2. Have the subject seated with his/her arm resting on a table. Wrap the pressure cuff snugly around the bare upper arm, making certain that the inflatable bag within the cuff is placed over the inside of the arm where it can exert pressure on the brachial artery. Wrap the end of the cuff around the arm and tuck it into the last turn, or press the fasteners together to secure the cuff on the arm. Close the valve on the bulb by turning it clockwise.
3. We will use the **auscultatory method** to indirectly measure blood pressure. In the auscultatory method the pressure cuff and a stethoscope are used to listen to changes in sounds in the brachial artery. Place the stethoscope

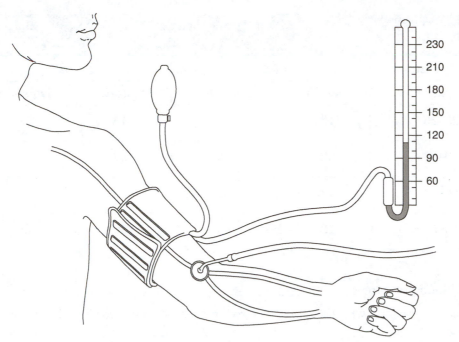

Figure 11.1 Technique for measuring blood pressure.

below the cuff and over the brachial artery where it branches into the radial and ulnar arteries (see the diagram). With no air in the cuff no sounds can be heard. Inflate the cuff so the pressure is above diastolic (80–90 mm Hg), and you will be able to hear the spurting of blood through the partially **occluded** artery.

4. Increase the cuff pressure to around 160 mm Hg; do NOT leave cuff pressure above 160 mm Hg for more than a few seconds. This should be above systolic pressure so that the artery is completely collapsed and no sounds are heard. Now, open the valve and begin to slowly lower the pressure in the cuff.

5. As the pressure is slowly allowed to decrease, you should be able to hear four phases of sound changes. These were first reported by Korotkov (in 1905) and hence are called Korotkov sounds:

    Phase 1—Appearance of a fairly sharp thudding sound which increases in intensity during the next 10 mm Hg of pressure drop. The pressure when the sound first appears is the **systolic pressure.**

    Phase 2—The sounds become a softer murmur during the next 10–15 mm Hg drop in pressure.

    Phase 3—The sounds become louder again and have a sharper thudding quality during the next 10–15 mm Hg of pressure drop.

    Phase 4—The sound suddenly becomes muffled and reduced in intensity. The pressure at this point is termed the **diastolic pressure.** This muffled sound continues for another 5 mm Hg pressure drop, after which all sounds disappear. The point where the sound ceases completely is called the **end diastolic pressure.** It is sometimes recorded along with the systolic and diastolic pressures in this manner: 120/80/75.

The auscultatory method has been found to be fairly close to the direct method in the pressures recorded; usually the actual systolic pressure is about 3–4 mm lower than that obtained with the direct method.

Table 11.1 Average blood pressure for female and male humans at various ages.

Age Range	Males		Females	
Years	Systolic	Diastolic	Systolic	Diastolic
20–29	124	77	116	73
30–39	126	79	122	76
40–49	129	81	128	81
50–59	136	83	138	84

My blood pressure is _____.

Check your value with the Table 11.1 giving average values for females and males of various ages.

How do your systolic and diastolic pressures compare to the average values for individuals of your age and sex? _____

Blood pressure varies with a person's age, weight, and sex. Below the age of 35 women generally have a blood pressure 10 mm Hg lower than that for men. However, after 40–45 years of age, a women's blood pressure increases more rapidly. The old rule of thumb of 100 plus your age is still a good estimate of what your systolic pressure should be at any given age. After the age of 50, however, the rule is invalid. The increase in blood pressure with age is caused largely by the overall loss of vessel elasticity with age, part of which is due to the increased deposition of **cholesterol** and other lipid materials in the blood vessel walls.

Although blood pressure measures how hard your heart is working, it tells a physician more about the health of your blood vessels. Patients are said to have high blood pressure or **hypertension,** when the resting systolic pressure is usually above 145 mm Hg and/or when the resting diastolic pressure is commonly above 90 mm Hg. These high pressures are usually the result of fatty deposits lying in the opening of the arteries, a medical condition called **atheriosclerosis.** Calcification of these deposits produces, **arteriosclerosis** or "hardening of the arteries." People with arteriosclerosis commonly have systolic pressures greater than 200 mm Hg and have **far** greater probabilities of suffering **heart attacks**, **aneurisms,** and **strokes** than people with normal blood pressures. An aneurism is the swelling of a weakened blood vessel, usually an artery, which can burst causing a stroke.

V. ELECTROCARDIOGRAM

Contractions of the atria and ventricles of your heart are triggered and hence coordinated by electrical stimuli. The electrical activity originates in a relatively small patch of tissue buried in the right atrium called the **sino-atrial node** (abbreviated SA node) or the **pacemaker.** The pacemaker is a unique tissue in that it possesses **automaticity,** i.e., it generates an electrical impulse spontaneously. The rate of impulse generation corresponds to a person's resting heart rate. (Heart rate is also influenced by two branches of the autonomic nervous system.)

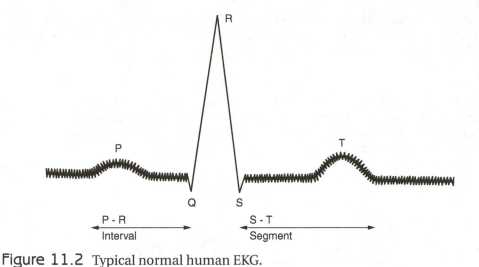

Figure 11.2 Typical normal human EKG.

In the normal healthy heart the electrical impulse generated at the SA node, travels first across the atria, through the interventricular septum and finally to and through both ventricles. This impulse pathway causes the heart to contract in an organized, coordinated fashion. If the impulse does not follow the normal pathway, medical problems will usually be apparent.

Physicians use an **electrocardiogram** mainly to noninvasively examine the electrical conduction through the heart. The diagram in Figure 11.2 represents the normal electrocardiogram (EKG). Each peak, wave and dip of the EKG is correlated with a particular electrical event within the heart.

Cardiologists have incorporated the letters P, Q, R, S, and T to identify various aspects of the EKG:

1. **P wave**—Initial impulse originating in SA node. A smooth p wave suggests the stimulus began in the SA node. Impulses arising from other locations are termed **ectopic.**
2. **P–R interval**—Indicates the time required for the impulse to travel from the SA node across the atria to the ventricles. Should not exceed 2 seconds.
3. **QRS complex**—Large scale electrical events accompanying contractions of the ventricles. If longer than about 0.12 second, conduction problems in ventricles are likely.
4. **T wave**—Represents the recovery phase of the ventricles after a complete contraction. People with severe heart damage often have an inverted T wave. Also, the length of time between S and T (S–T segment) in coronary patients is usually longer because damaged heart muscles require more time to recover.

The normal EKG consists of a successive series of electrical events represented by the letters P, Q, R, S and T about 60 to 80 times per minute in resting, healthy people.

Many conditions can be diagnosed from EKG records including arrhythmias, bradycardia, and tachycardia which were previously discussed. Some others include impulses arising from areas other than the pacemaker **(ectopic),** fluttering beats, and **fibrillation.** The latter is the most critical condition. The electrical activity more or less gets out of hand. In atrial flutter, the atria contracts some 250 to 400 times per minute. These contractions are shallow and uncontrolled. If the ventricles "get into" fibrillation, the complete EKG pattern is lost and hence total disorganization occurs. Fibrillation is not compatible with life.

Procedure

It will not be possible to record an EKG for each student in the laboratory. Your instructor will call for volunteers, hopefully representing both fit and unfit categories. The results from these individuals will be discussed by the entire class. If no EKG's are available from the class, use the examples below.

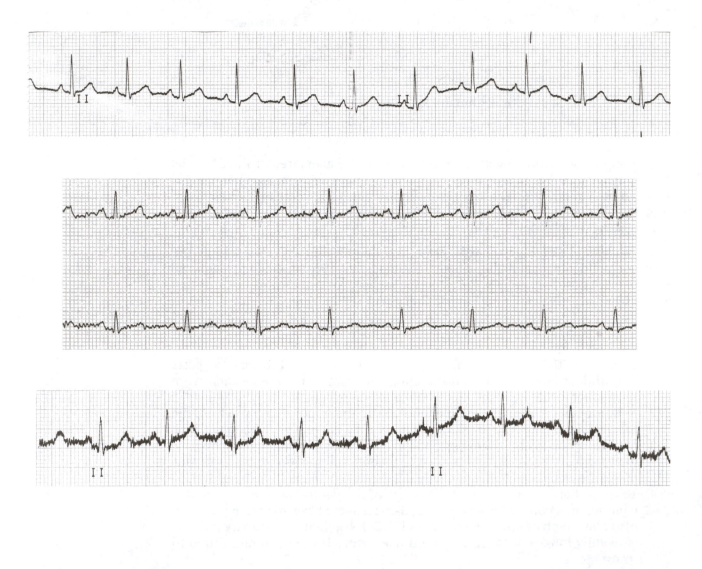

REVIEW QUESTIONS

I. On part A below, carefully draw in three PQRST events of the normal EKG. Make the R "spike" coincide with each of the vertical lines marked R. In part B, indicate an EKG for a person whose heart is beating twice as fast as the heart in part A.

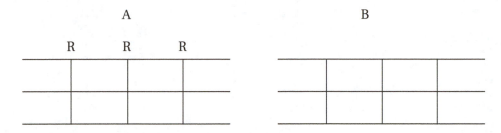

II. Please provide the correct answer to the following questions.

    1. Slower than normal heart rate.

    2. Site where "heartbeat" originates.

    3. Higher than normal blood pressure.

    4. Highest of the two blood pressure measurements.

    5. Total disorganization of heart contractions.

    6. Calcified deposits within the arteries.

    7. Unequal time between "heart beats."

    8. Unusual sounds from within the heart.

    9. Device for measuring blood pressure.

    10. Medical name for a heart attack.

III. Please provide the following data.

    1. My heart rate was _____.

    2. My blood pressure was _____ (if available).

    3. If either are considerably higher or lower than the population norms, do you have any explanation?

    _____

Cardiopulmonary Anatomy and Physiology

I. INTRODUCTION

Each of your cells is a single living organism. It's basic needs are oxygen and nutrients in the form of amino acids, sugars, lipids, nucleotides and a few other things like vitamins and minerals. It also produces waste products, which if allowed to accumulate threaten the cell's survival. Free living cells must move to hunt down nutrients and flee their own wastes, but a cell in your body doesn't have that option. Without some specialized system replenishing locally depleted nutrients and washing away cellular waste products, cells would quickly die, in some cases in the brain and heart within minutes.

The "specialized system" that provides these services, of course, is the **cardiovascular system**—the heart, blood vessels, and blood. Blood brings nutrients and oxygen required to power cellular metabolism and carries away carbon dioxide, urea and other waste products. In its circuit, blood also visits certain "blood conditioning organs," like lungs and kidneys, that remove wastes from the body and replenish or conserve nutrients. But that's not all blood does. It also buffers the acid-base balance of the body, destroys foreign organisms through phagocytosis and antibody action, distributes and conserves body heat and prevents its own loss through mechanisms of hemostasis (blood coagulation and clotting), among other things.

Our goal in this laboratory is to examine the interrelationships of the cardiovascular system and a key blood conditioning system, the **respiratory system.** This system supplies oxygen to the blood for distribution around the body and also removes carbon dioxide from the body. Both services support the same function—**cellular respiration**. To manufacture sufficient **adenosine triphosphate**, cells need oxygen to facilitate the electron transport system, which provides the energy for oxidative phosphorylation of ADP to make ATP. But, the Krebs (citric acid) cycle, a biochemical pathway that precedes the electron transport chain, produces carbon dioxide as a byproduct. If this carbon dioxide accumulates, it causes the cell's internal and external environments to become acidic, which in turn alters the function of key enzymes, ultimately threatening the cell's ability to maintain *homeostasis,* a constant internal environment. So, in combination these two systems, cardiovascular and respiratory, are absolutely critical for survival.

II. HEART ANATOMY

The primary function of the heart is to move blood. It does so by pumping blood through a series of four chambers in a characteristic pattern. Each chamber is surrounded by muscle that, when it contracts, squeezes the chamber like a balloon, which forces blood out through a hole. The four chambers are the following (consult your textbook and posters in lab for appropriate pictures):

1. **Right Atrium:** Thin-walled chamber in the upper right portion of the heart. Pumps blood into the right ventricle.
2. **Right Ventricle:** Thicker walled chamber in the lower right portion of the heart. Pumps blood towards the lungs.
3. **Left Atrium:** Thin-walled chamber in upper left portion of the heart. Pumps blood into the left ventricle.
4. **Left Ventricle:** Strongest portion of the heart and has the thickest walls. Pumps blood throughout the body.

Notice that the ventricles pump blood out of the heart. Each ventricle forces blood into an **artery,** which by definition is **a blood vessel carrying blood away from the heart.** The **atria** (plural of atrium), on the other hand, simply pump blood into the ventricles. They receive blood returning from the body from large **veins.** A **vein** is **a blood vessel that carries blood towards the heart.** Therefore, the atria receive blood from veins, and the ventricles pump blood into arteries. Here are the details:

1. **Right Atrium:** Receives blood from superior and inferior vena cava.
2. **Right Ventricle:** Pumps blood into the pulmonary trunk.
3. **Left Atrium:** Receives blood from the pulmonary veins.
4. **Left Ventricle:** Pumps blood into the aorta.

A series of veins returns blood to the heart from specific regions of the body:

1. **Superior vena cava:** Drains blood from the head and arms.
2. **Inferior vena cava:** Drains blood from the lower portion of the body.
3. **Pulmonary veins:** Drains blood from the lungs.

Similarly, the arteries carry blood to specific regions:

1. **Pulmonary trunk:** Brings blood to the lungs.
2. **Aorta:** Carries blood to the entire body.

After the ventricles have filled with blood they contract and push the blood out through their respective arteries much like squeezing an untied water balloon will push water out its top. However, there is a small problem. Not only is there a hole between the ventricle and the artery, but there is also a hole between the ventricle and its atrium. We don't want blood to squirt back into the atrium; it just came from there. We want it to be pushed into the artery. So, there is a pair of one-way valves between atria and ventricles that keep blood flowing in the proper direction. Also, when the ventricle relaxes, we don't want the blood we just pumped into the artery to gush back into the ventricle, so there is a second set of valves between ventricles and their respective arteries. Each of these valves has a name:

1. **Tricuspid valve:** Between the right atrium and right ventricle. Will not allow blood to be pumped from the ventricle into the atrium.
2. **Bicuspid (Mitral) valve:** Between the left atrium and left ventricle. Will not allow blood to be pumped from the ventricle into the atrium.
3. **Pulmonary semilunar valve:** Between the right ventricle and pulmonary trunk. Will not allow blood to flow from the trunk back into the ventricle.
4. **Aortic semilunar valve:** Between the left ventricle and aorta. Will not allow blood to flow from the aorta into the ventricle.

The general circulatory pattern followed by blood is:

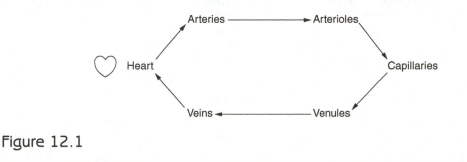

Figure 12.1

III. CIRCULATORY SYSTEM IN HUMANS

Mammals have a four-chambered heart (two atria and two ventricles) which provides complete separation of venous and arterial blood.

In reality, the human heart consists of two separate but joined hearts. The **right atrium** collects venous blood via the two large **vena cavas** (posterior which drains the lower parts of the body and the anterior which returns venous blood from the head) and passes it to the **right ventricle** which pumps this blood via the **pulmonary arteries** to the **lungs**. After passing through the lungs, oxygenated blood returns to the **left atrium** via the **pulmonary veins**. It then enters the **left ventricle** where it is pumped out of the heart via the **aorta**. Consequently, we have two complete and separated circuits in mammals: the **pulmonary circuit** to and from the lungs which is responsible for oxygenating blood and the **systemic circuit** which distributes oxygenated blood to the entire body. A single aortic arch (the left one in mammals) gives rise to the dorsal aorta. The dorsal aorta is the main distribution artery in the body.

The **pulmonary trunk** bifurcates into the pulmonary arteries. The pulmonary arteries from the right ventricle conduct deoxygenated blood to the lungs. The **pulmonary veins** return oxygenated blood to the left atrium in mammals.

IV. HEART PHYSIOLOGY

Exercising causes a person to "burn" ATP to power the working muscles. Since many muscle cells (not all) replenish ATP using oxidative phosphorylation, their requirement for oxygen increases as does their need for carbon dioxide removal. Therefore, working muscles demand more blood. The heart accommodates by increasing its rate (the number of beats per minute) and the amount of blood it pumps in one minute, called the **stroke volume.**

Table 12.1 Cardiac data for subjects at rest.

Initials	Smoker?	Heart rate (bpm)	Stroke volume (ml)	Cardiac output
			70	
			70	
			70	
			70	

Table 12.2 Cardiac data for subjects after exercise.

Initials	Heart rate (bpm)	Stroke volume (ml)	Cardiac output	Difference
		140		
		140		
		140		
		140		

If we know both stroke volume and heart rate then we can calculate the amount of blood the heart pumps in one minute, a quantity called the **cardiac output.** In particular, if we let Q be cardiac output, SV be stroke volume and HR be heart rate, then

$$Q = SV \times HR.$$

In the following experiment we will measure change during exercise in cardiac output for various members of the class and compare smokers versus nonsmokers.

1. Participation in the following exercises are strictly voluntary. *If you are pregnant or have a medical condition that makes exercise at all risky, you may not be the experimental subject.*
2. Start by determining the resting heart rate of all participating members of the group. Find the pulse at the wrist on the thumb side or at the base of the neck. Place the data for your group members in Table 12.1.
3. Calculate cardiac output from these data assuming that each person's stroke volume is 70 ml.
4. Anyone who wants to should now begin exercising. Options include jumping jacks, sit-ups, push-ups or any other *safe* creative exercise that will increase heart rate. The moment a group member finishes exercising, measure heart rate and record the results in Table 12.2.
5. Stroke volume typically doubles during exercise, so we will assume that is the case for every member of the group who exercised. With that assumption, calculate the cardiac output after exercise.

6. In the final column of Table 12.2 calculate the difference in cardiac output before and after exercise. In particular, calculate

$$Q_{diff} = Q_{exer} - Q_{rest,}$$

where Q_{diff} is the difference in cardiac output, Q_{exer} is cardiac output after exercise and Q_{rest} is resting cardiac output.

On the figure provided identify the following structures (use your textbook and other resources in the lab to help you learn them):

Right and left atria	Interventricular septum
Right and left ventricles	Aorta
Mitral or bicuspid valve	Superior and inferior vena cava
Tricuspid valve	Aortic and pulmonary semilunar valves
Pulmonary trunk	Pulmonary arteries
Pulmonary veins	

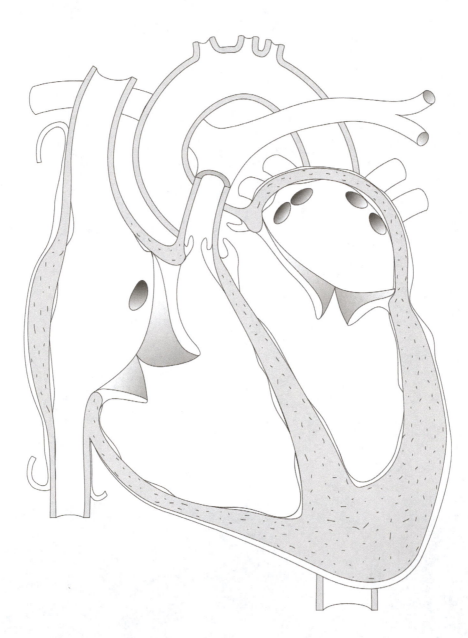

Figure 12.2   Sectioned heart. © 2003 Mark Nielsen. Art by Jamey Garbett.

In fourteen steps follow the flow of blood through the heart. Begin with the Superior Vena Cava and end with the aorta. Don't forget to put in all the valves.

1. Superior Vena Cava

2.

3.

4.

5.

6.

7.

8. Lungs

9.

10.

11.

12.

13.

14. Aorta

REVIEW QUESTIONS

I.   Please provide the correct answers to the following questions.

1. A vein that carries oxygenated blood.

2. The largest artery in your body.

3. Storage chambers of the heart.

4. Vessels where exchanges of gases and materials occur.

5. Organ connecting pulmonary and systemic circulatory systems.

6. Structures separating atria and ventricles.

7. Area served by a subclavian artery.

8. Artery carrying blood to the heart.

9. Circuit delivering oxygenated arterial blood to body.

10. Major vein returning venous blood from the head region to the heart.

V. RESPIRATORY SYSTEM ANATOMY

The respiratory system consists of the lungs and air passages that conduct air from outside the body into the lungs. In addition one could include muscles that expand and contract the chest. It's this expansion and contraction of chest volume that moves air into and out of the lungs, a process called **ventilation.**

1. On the mannequin model and posters find and prepare to identify on a practical examination the following structures (use your textbook and other resources in the lab to help you learn them):

Lung	Primary bronchus
Trachea	Secondary bronchus
Larynx	Nasal cavity
Pharynx	Palate

2. Find the diaphragm muscle on the mannequin model. When it contracts, what happens to the volume of the thoracic cavity? (Look carefully at the anatomy and think about what would happen to the shape of the muscle when it contracts.)

3. Examine the chest cavity model (half a bell jar with a rubber sheet on the bottom) in the supply area. The jar represents the chest cavity. What are the following structures analogous to?

   Balloons   _____

   Tubes connected to balloons   _____

   Rubber sheet/handle   _____

4. Gently pull the sheet down a few times or press the top handle and observe what happens to the balloons. Answer the following questions.

   (a) Does pulling the balloons down represent contraction or relaxation of the diaphragm?

   (b) What happens to the balloons?

   (c) What happens to the pressure inside the jar when you pull the rubber sheet down and how do you know?

5. What is the function of the internal and external intercostal muscles? Where are these muscles located? (Use your text and other resources as references.)

VI. RESPIRATORY PHYSIOLOGY

There are a number of important measurements that can be made to test lung function. One of the most important is **vital capacity,** which is the total amount of air one can move in a single maximal breath. To make the measurement, the subject breathes in as much air as they can hold and then releases it smoothly into a device called a **spirometer.** The spirometer measures how much air flows into it. The objective in this portion of the exercise is to measure vital capacities of all students and determine how these measurements vary by height and gender.

1. Each student should obtain their own clean spirometer mouthpiece to use throughout the entire lab. Please discard the mouthpiece into the trash when finished.
2. Follow the same procedures for each member of the group.
   (a) Make sure the spirometer is set to zero before beginning.
   (b) Take a few deep warm-up breaths. Then take as deep a breath as you can and exhale as much as possible into the spirometer. Breathe out smoothly and under control; you do not have to push air out forcefully.
   (c) Record your tidal volume on a separate sheet of paper.
   (d) Reset the spirometer to zero and repeat the measurement twice more.
   (e) Average the results and place your data in Table 12.3.
3. Record heights of all group members *in inches* in the proper location of Table 12.3.

Table 12.3 Group vital capacity data.				
Initials	*Gender*	*Smoker?*	*Ave. Tidal Volume*	*Height*
	140			
	140			
	140			
	140			

4. On the graph paper below draw a scatter graph of the tidal volume data. Place the independent variable on the horizontal axis and the dependent variable along the vertical. Part of the exercise is for you to determine what the variables are and which are independent and dependent. ***Don't forget to label the axes.***

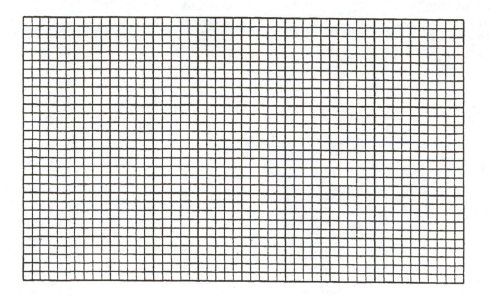

CARDIOPULMONARY REVIEW FIGURES AND QUESTIONS

On the figures provided identify the following structures (use your textbook and other resources in the lab).

Larynx
Right Lung
Trachea
Primary bronchus
Left lung

Tracheal cartilage
Alveolus
Arteriole
Capillary
Venuole

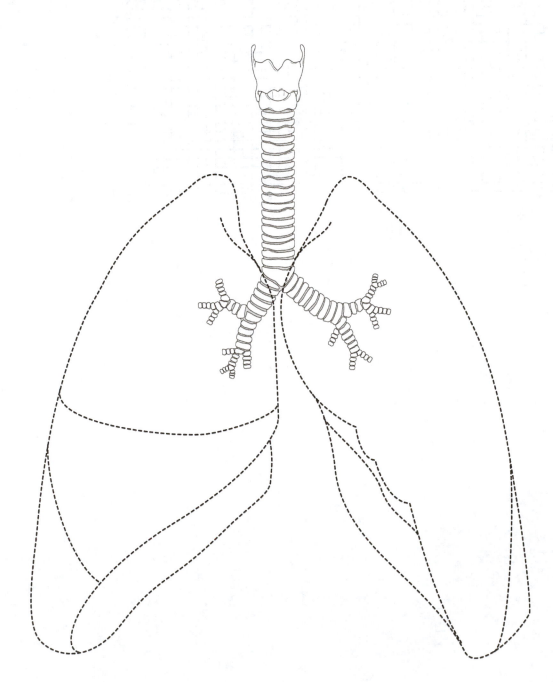

Figure 12.3 Trachea, cross section. © 2003 Mark Nielsen. Art by Jamey Garbett.

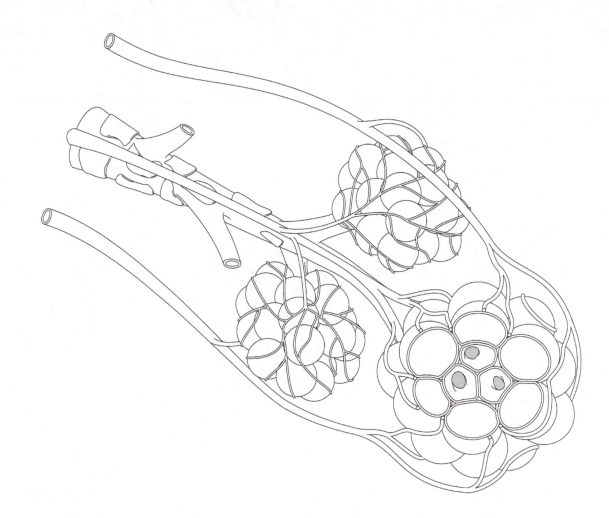

Figure 12.4 Microstructure of lungs. © 2003 Mark Nielsen. Art by Jamey Garbett.

Excretory System

I. INTRODUCTION

The kidneys play a vital role in homeostasis, the maintenance of a constant internal environment. As filtering devices in your body, the kidneys regulate the chemical content of the blood, including parameters such as pH, salt content, glucose content, and water balance. While you may have pictured the primary function of the kidneys to be for excreting water, these other regulatory functions are also important.

The kidneys form urine by the process of **filtration** and something known as **reabsorption**. Water and many chemical substances of small molecular size are filtered out of the blood through small capillaries into a hollow bulb with very thin cell layers. In order for the chemical substances to pass through the capillaries into the kidney tubular system, a great deal of water must be filtered, as much as 180 liters (190 quarts!!) in a day. Surely you do not lose that much water in a day, or you would not be able to sit in lab for more than 10 minutes without urinating. What happens to all of the water? Much of the water is reabsorbed back into the blood stream through special structures of the kidneys. Some chemical substances that are passively filtered also go back into the blood. The result is that a great deal of water is initially passed into the kidney tubule system and then removed so that the resultant urine becomes concentrated. You will learn more about this later.

We will begin our study of the excretory system by examining the gross anatomy of the urinary tract. We will then examine the microscopic structure of the kidney using slides and models. Finally, we will gain an understanding of how the kidneys deal with excess of certain substances by conducting simple experiments on fluid and solute intake and urine analysis.

READ CAREFULLY!! In order to get everything done in this exercise, you must be organized. Begin the urine experiment first, and then proceed through the examination of the kidney.

II. URINE ANALYSIS

Each student should urinate, collect the urine, and measure its pH, volume, specific gravity, and glucose content using the methods described. Students should perform each of these measurements or tests on the urine sample collected at the beginning of the experiment. Record the data in the table provided.

Urine pH Determination—Check the pH of each urine sample using pH indicator paper. Dip the indicator paper in the urine sample and compare the paper's color to the indicator color chart by holding it up to the light. Record the pH in the table.

From *Animal & Human Biology Exercises* by Zimmerman et al. Copyright © 1991 by Kendall/Hunt Publishing Company. Used with permission.

Volume Measurement—Use a graduated cylinder to measure the volume of the urine sample and record the volume in the table.

Specific Gravity Measurement—Use a hydrometer, also known as a urinometer, to measure the specific gravity of the urine samples. Pour the sample to be tested into the graduated cylinder. There must be sufficient sample for the hydrometer to float. Be sure the hydrometer does not contact the side of the cylinder. Note the point at which the meniscus of the urine intersects the hydrometer scale. The numbers on the scale represent a specific gravity of 1.00 or higher. Only the last two digits of the reading may be seen at some points on the scale. For example, if the meniscus intersects the scale at 30, the specific gravity should be recorded as 1.030. Hydrometers are calibrated at 15°C. Check the temperature of the urine sample, and add 0.001 to the specific gravity for every 3° above 15°C or subtract 0.001 for every 3° below 15°C.

Estimation of Urinary Solids—If a sample of urine were evaporated, the solids contained could be recovered and weighed. These solids would include organic and inorganic compounds excreted by your body. There is another way to estimate this figure. It can be computed by multiplying the last two digits of the specific gravity by 2.66, a value known as Long's constant. For example, if the specific gravity of a urine sample were 1.030, the estimated concentration of solids would be 30 × 2.66 = 79.8 g/liter.

Sugar Tests—To test for glucose concentration, dip a Clintest strip in the urine sample for about 15–20 sec. The test strip should change color. Now compare the color of the test strip to the color chart. Record the concentration of glucose in the table. Note: Two kinds of test strips are available, some will give only comparative concentrations, i.e., negative, low, medium, or high while others will give numerical concentrations.

If another test strip is available it will be possible to test for leukocytes, nitrite, protein, pH, blood, specific gravity and ketones as well as glucose.

Table 13.1

Vol. ml	pH	Specific Gravity	Glucose	Urinary Solids	WBC	Nitrite	Protein	Blood	Ketones

From *Animal & Human Biology Exercises* by Zimmerman et al. Copyright © 1991 by Kendall/Hunt Publishing Company. Used with permission.

III. ANATOMY OF THE URINARY SYSTEM

The **kidneys** are covered and attached to the dorsal body wall by a tough membrane, the **peritoneum**. The urinary tract is comprised of two elongate kidneys with a narrow band of white tissue, the **adrenal gland** on the surface. You should now note the following: several tubular structures enter or leave the kidney. Two of these are blood vessels associated with the **renal portal system** which serves to filter blood in the kidneys. Next, note the small, flaccid tube that runs along and leads from the kidney posteriorly. This is the **ureter**. A **urinary bladder** is present in the pelvic cavity. The **urethra** is a single tube used to void the bladder.

If available, you may now dissect a mammalian kidney. Using a sharp scalpel, make a longitudinal slice starting at the outer convex region of the kidney. Do this very carefully. Open the kidney halves, observe the following and compare with Figure 13.1; the **renal cortex**, the outer one-third of the kidney; the **renal medulla**, the middle one-third of the kidney; and the **renal pelvis** to which the ureter is attached. These structures will be put into perspective through examination of microscopic sections of kidneys and the figure of the nephron and models.

IV. MICROSCOPIC EXAMINATION OF THE KIDNEY

First, examine the drawing of the **nephron** in Figure 13.2. You should observe carefully as you trace the path of blood and urine. Blood that is to be filtered passes into the kidney via the renal artery. This large artery breaks into smaller arterioles which end in globular masses of capillaries, the **glomeruli** (singular: **glomerulus**). Surrounding each of the glomeruli are double-walled filtering structures known as **Bowman's capsules**. Dissolved substances to be filtered from the blood diffuse from the thin-walled capillaries of the glomerulus into the Bowman's capsules. In order to facilitate the passage of these dissolved substances (sugars, amino acids, salts, hormones, etc.) a great deal of water is initially filtered in order to "wash" the materials through the tubules of the kidneys. From the Bowman's capsule, the dilute **glomerular filtrate** (urine) passes first through the **proximal convoluted tubule**, then the **Loop of Henle**, and then the **distal convoluted tubule**. The dilute urine becomes concentrated as it passes through this tubular system, as water diffuses back into capillaries which surround convoluted tubules and, especially, the Loop of Henle. Finally, this concentrate leaves the tubular system via the **collecting tubules** and passes to the pelvis of the kidney, where it leaves the kidney by way of the ureter.

Now refer back to the dissected kidney. The cortex of the kidney is comprised of glomeruli, Bowman's capsules, and convoluted tubules. The renal medulla is comprised of Loops of Henle and collecting tubules. The larger collecting tubules merge toward the renal pelvis; the merging of several of these appears as a **pyramid** with its apex at the **pelvis**.

You should now examine slides of renal tissue, noting the numerous glomeruli and Bowman's capsules in the cortex and the tubules of the medullary region.

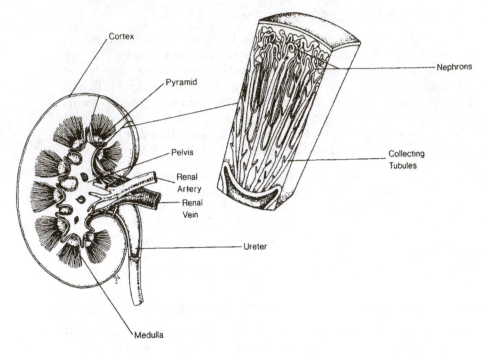

Figure 13.1   Longitudinal section of human kidney.

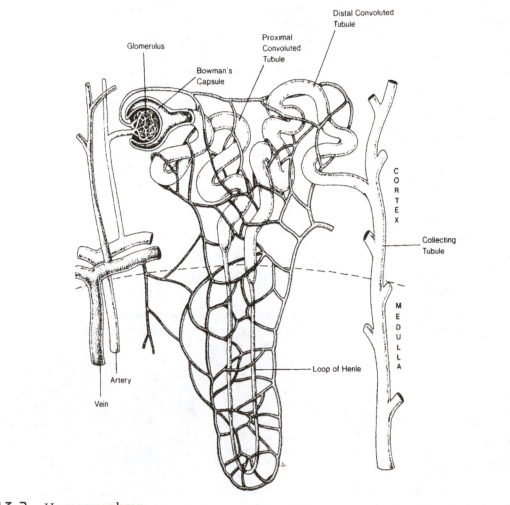

Figure 13.2   Human nephron.

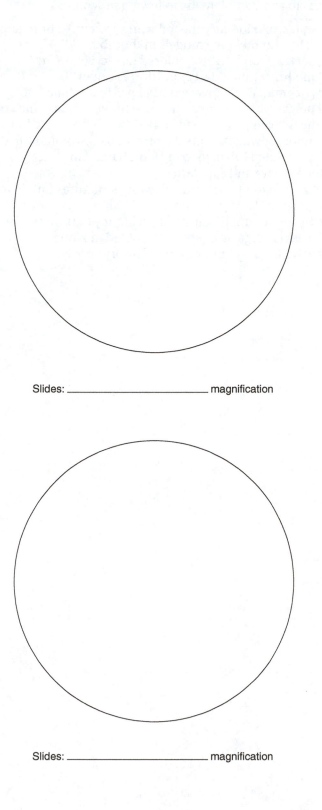

Slides: _____ magnification

Slides: _____ magnification

V. RENAL ANATOMY AND HISTOLOGY

During periods of relative inactivity, apply yourself to the anatomy models, posters and texts in the lab and complete the following procedures.

1. Locate and learn to identify the following structures or regions on the following figure (13.3) and kidney models in the lab:
    Renal cortex: Outer, lighter-colored portion of organ
    Renal medulla: Inner, darker portion of organ
    Renal pyramid: Dark, inverted triangles in medulla
    Renal pelvis: Chamber of connective tissue where urine collects
2. Identify the following on figure (13.4):
    Glumerolus, Bowman's capsule, proximal convoluted tubal,
    distal convoluted tubal, loop of Henle, collecting duct,
    bladder, kidneys and capillaries.
3. Locate and learn to identify the following structures on the torso mannequin and on the figure (13.5) provided:
    Ureter: Tube conducting urine from kidney to urinary bladder
    Renal artery: Large artery bringing blood to kidney
    Renal vein: Large vein next to renal artery

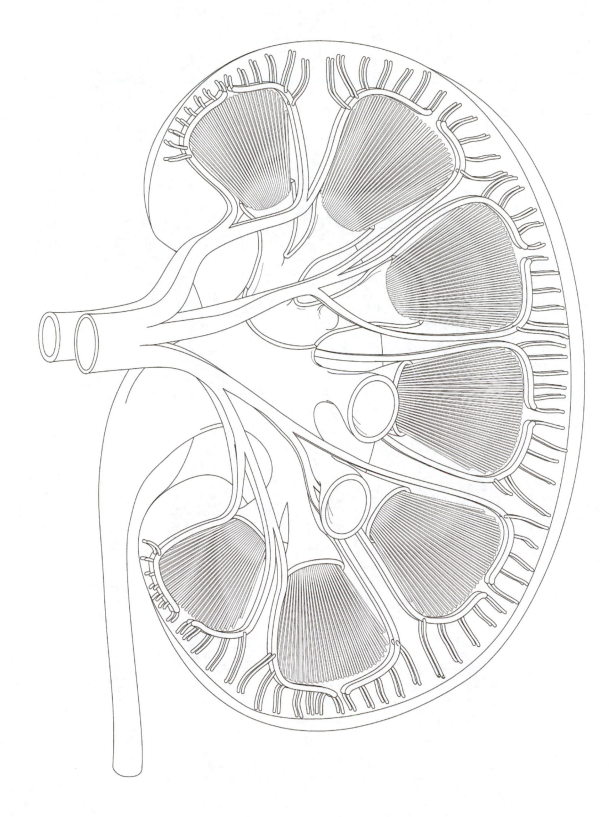

Figure 13.3   Vessels of the kidney. © 2003 Mark Nielsen. Art by Jamey Garbett.

Figure 13.4 Microstructure of the kidney. © 2003 Mark Nielsen. Art by Jamey Garbett.

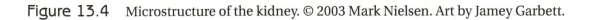

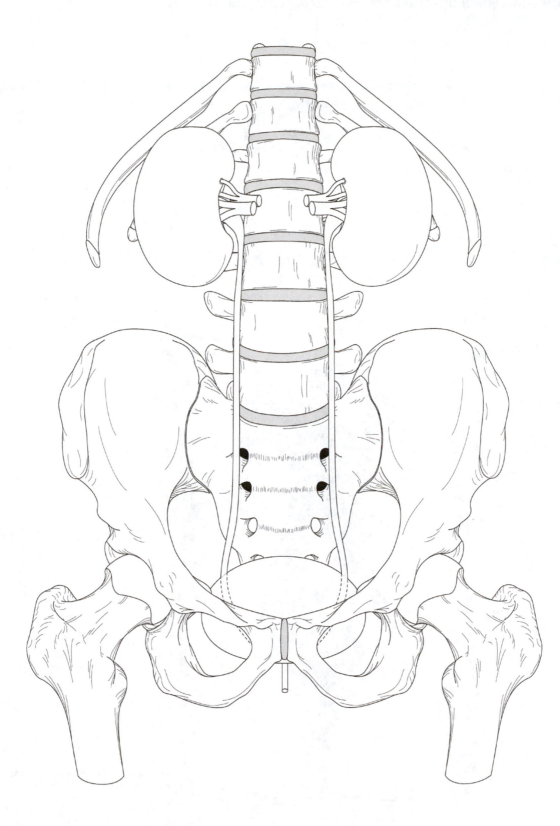

Figure 13.5　Urinary System. © 2003 Mark Nielsen. Art by Jamey Garbett.

Identify the parts of the urinary system. Also identify the diaphragm, liver, stomach, spleen, large intestine, small intestine, rectum and adrenal glands.

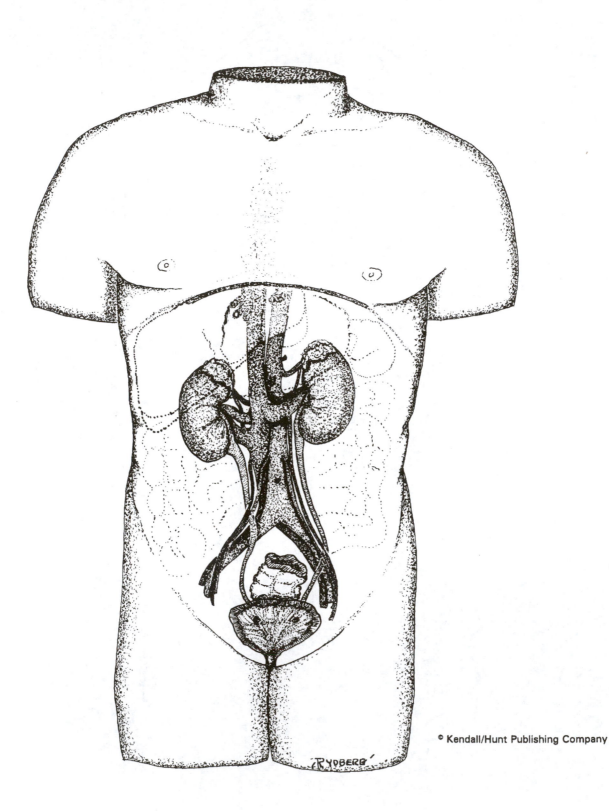

REVIEW QUESTIONS

1. Diagram and label a nephron in the space below:

2. What is the function of each of the following portions of the nephron?
   (a) Bowman's capsule:

   (b) Proximal convoluted tubule:

   (c) Loop of Henle:

   (d) Distal convoluted tubule:

   (e) Collecting duct:

3. What epithelial cell types are found in the urinary system? Explain their benefit to the urinary system.

NOTE: Start keeping the Dietary Log from Exercise 15 to use next lab period.

Digestion

I. INTRODUCTION

Every cell in the body, from a four-foot long nerve cell to the smallest sperm cell, performs work. Work requires energy, and energy consumes fuel. Fuel for the body's energy comes from the food you eat—carbohydrates, proteins, and fats. Your food also provides raw materials for building, repairing, and controlling your body systems.

The digestive system is truly a remarkable machine. During a person's lifetime, it may process between 60,000 and 100,000 pounds of food.

Digestion breaks down food into nutrients—basic materials the body can use. This process takes place chiefly in the **alimentary canal**, a long tube beginning at the mouth and ending at the anus. In adults the canal is about 27 feet long. The liver and pancreas aid in the work of digestion.

The mechanical breakdown of food occurs with the chewing process. In order for food to be absorbed into the bloodstream, it must be broken down into molecules that are small enough to diffuse through cell membranes. This chemical breakdown of food is accomplished by the action of digestive enzymes, water and emulsifying agents on the food molecules. **Digestion** is a series of chemical reactions occurring in the mouth, stomach, and small intestine, in which large food molecules are broken down into smaller food molecules. The end products of digestion are absorbed through the cells that line the inside of the intestine and enter the blood vessels in the intestinal walls.

II. PROCEDURES

Absorption of Glucose Through the Intestinal Walls

Glucose is absorbed through the walls of the intestine by **diffusion**. Diffusion is the movement of a substance (other than water) through a differentially permeable membrane from a region of high concentration to a region of lower concentration. A differentially permeable membrane is a membrane that only allows a certain substance through, usually because of the size of the molecules of the substance. If the molecules are too big, they can't pass through. If the molecules are small enough, they can pass.

You will use dialysis tubing to simulate the intestinal wall.

1. Obtain an 8″ piece of dialysis tubing which has been soaking in water. Open the tubing to make a column. Tie one end with thread so that it does not leak.

2. Fill your tied tubing about 1/3 full with glucose.
3. Next add starch solution in the sack until it is about 2/3 full. Tie off the open end. RINSE THE TIED SACK WITH TAP WATER.
4. Place your sack into a 150 ml beaker. Cover it with water to the 100 ml mark. Let the sack sit in the beaker for 30 minutes or more while you do the rest of the lab exercise. You may gently bend and fold the sack to simulate the peristaltic movement of the small intestine.

The Action of Bile on Fats

Bile is a substance produced by the liver and secreted through the bile duct into the small intestine. Bile is **not** a digestive enzyme. The function of bile is to **emulsify fat** (it breaks up large blobs of fat into smaller blobs of fat). For this reason, bile is sometimes called an emulsifying agent. Fat is not digested by the action of bile, but only reduced to smaller blobs. These blobs, however, are still too large to be absorbed into the wall of the intestine. The surface area of these smaller blobs is **much** greater than the surface area of the original **large** blobs of fat and, therefore, it is easier and faster for digestive enzymes to **hydrolyze** (digest) the fat into smaller molecules that **can** be absorbed into the wall of the intestine.

1. Obtain two test tubes and label them A and B. Be **certain** the tubes have been washed and are **clean.** With a marker make lines 1 cm and 2 cm from the bottom of the tubes.
2. Place cooking oil (fat) in test tube A to the 1 cm mark. Add water to the 2 cm mark and 6 drops of Sudan Red (a stain that "clings" to fat droplets so they will show up better).
3. Cap the test tube and shake vigorously for 5 minutes. Set the tube in your test tube rack.
4. Place cooking oil (fat) in test tube B to 1 cm mark. Add water to 2 cm mark, 6 drops of Sudan Red, and one drop of liquid soap. Cap the tube and shake vigorously for 5 minutes (in the intestines, this "shaking" would be accomplished by muscular contractions of the intestines). Set the tube in your test tube rack. Liquid soap is also an emulsifying agent and will do the same thing to fat in a test tube that bile will do to fat in the intestines.
5. Wait 2 minutes and observe tubes A and B. Specifically, observe the area below the red band.

Table 14.1 Results of Sudan Red Test		
Test Tube	*Contents*	*Results*
A	Oil, water	
B	Oil, water, detergent	

The Action of Lipase on Fats

Lipase is a digestive enzyme that will hydrolyze (digest) fat molecules into molecules of **glycerol** and **fatty acids.** Gycerol and fatty acid molecules are small enough to be absorbed into the wall of the intestine.

1. Obtain two test tubes and label them A and B. Be **certain** the tubes have been washed and are **clean.** Mark these tubes at 2 cm, 4 cm, and 7 cm.
2. Pour fat (cream) in each tube to the 2 cm mark.

3. Add 1% litmus solution to **each** tube to the 4 cm mark. (**Acids** will turn litmus to a **pink** color). Shake the tubes to mix the fat (cream) and litmus.
4. Add the enzyme **lipase** to the 7 cm mark in tube A only. Shake tube A again.
5. Place tubes A and B in a water bath adjusted to 37° C (body temperature).
6. Wait 5 minutes and watch for a change in color in tubes A and B.

Table 14.2	Results of Litmus Test	
Test Tube	***Contents***	***Results***
A	Cream, lipase	
B	Cream	

Digestion in the Mouth

Salivary amylase is a digestive enzyme secreted into the mouth by the salivary glands. It will hydrolyze (digest) large **starch** molecules into smaller **maltose** molecules. Maltose molecules are still too large to be absorbed into the wall of the intestine. When maltose has been swallowed and passed into the intestine, another digestive enzyme will act on it and hydrolyze (digest) it into **glucose.** Glucose molecules are small enough to be absorbed into the wall of the intestine.

In this procedure, you will use an **artificial human saliva** (amylase). This is to eliminate the risk of infectious organisms in real salivary secretions. Real saliva works in exactly the same manner. Detection of the action of amylase, the enzyme in saliva, will be seen by observing the disappearance of starch. Iodine (IKI) solution is an indicator of the presence or absence of starch. Iodine reacts with starch to yield a blue-black substance. In the absence of starch, the iodine will remain yellowish-brown. You will also use Benedict's solution which is an indicator for sugar. If the amylase turned the starch to maltose, Benedict's solution will turn from blue to orange on boiling. If no sugar is present, Benedict's will remain blue on boiling.

1. Obtain four tubes and label them A, B, C, and D. Be **certain** the tubes have been washed and are **clean.**
2. **Place sufficient powdered** starch in test tube A to barely cover the bottom of the tube. Fill the tube 1/2 full of water. Cap the tube, and thoroughly mix the starch with the water by shaking the tube.
3. Add 5 drops of **iodine** to test tube A. A purple, blue, or black color indicates a positive test for the presence of starch. Set the tube in your test tube rack and save it for later.
4. Place sufficient **powdered** starch in test tube B to barely cover the bottom of the tube. Fill the tube 1/2 full of water. Cap the tube, and thoroughly mix the starch with the water just as you did in Step 2.
5. Add 10 drops of **Benedict's solution** to test tube B.
6. Shake up test tube B. Put it into a beaker of boiling water. Remove the cap from the top of the test tube. Let it boil for about 3 minutes. Remove the tube and set it in the rack to cool. If the mixture turns yellow, maltose is present.
7. To test tube C add powdered starch to cover the bottom, fill the tube 1/3 with water, and shake well. Next add 30 drops of artificial saliva. Shake again.
8. Place 1/2 of the mixture from test tube C into test tube D.
9. Add 10 drops of Benedict's solution to test tube D, shake, remove the cap, and place in boiling water. Look for color change.
10. To test tube C add 10 drops of iodine. Do you still have starch present?

Table 14.3	Results of Amylase Test	
Test Tube	*Contents*	*Presence of starch or maltose*
A	Starch, iodine	
B	Starch, Benedicts	
C	Starch, amylase, iodine	
D	Starch, amylase, Benedicts	

Absorption of Glucose in Your Artificial Intestine

Your sack has been sitting in water for about a half hour at this point. Now you will test to see if glucose and/or starch diffused through the membrane.

1. Obtain two clean test tubes and label them A and B. Put about 1 cm of water outside the sack in each of the test tubes.
2. Add 10 drops of iodine to tube A. Shake, if the solution turned black, you have starch present in the outside water.
3. Add 10 drops of Benedict's reagent to tube B. Put the tube into a beaker of boiling water. If the solution turns reddish, glucose is present in the outside water.

REVIEW QUESTIONS

1. Did glucose or starch diffuse through the walls of your artificial intestine?

2. Explain why starch must be digested to glucose based on the results of your intestine experiment.

3. In the simulated bile experiment, in which tube did the fat droplets appear larger (don't just give number but describe the tube)?

4. What is the advantage of having bile break large fat droplets into smaller droplets?

5. Why did you use litmus solution in the lipase experiment?

6. Can the fats (triglycerides) in cream be converted to fatty acids without lipase?

7. In the amylase experiment, was any of the starch changed (digested) to maltose? In which tube (don't just give number)?

8. For what does Benedict's Reagent test?

9. Where is bile produced?

From *Human Biology: A Laboratory Manual*, Fifth Edition/Revised Printing by Roberta B. Williams. Copyright © 2000 by Kendall/Hunt Publishing Company. Used with permission.

10. Where is amylase produced?

11. What are the four sphincters of the GI system?

12. Why are the liver, pancreas, gall bladder, and salivary glands accessory organs of digestion?

13. What are the epithelial cells of the alimentary canal?

14. If you put drops of salad oil and drops of water on a brown paper bag which will evaporate first? Why?

15. Identify the following organs on Figure 14.1: **mouth, parotid salivary gland, sublingual salivary gland, submandibular salivary gland, tongue, pharynx, esophagus, stomach, spleen, liver, gall bladder, pancreas, small intestine— a) duodenum b) jejunum c) illeum and large intestine—a) anus b) rectum c) sigmoid colon d) appendix e) cecum f) descending colon g) ascending colon h) transverse colon**

From *Human Biology Laboratory Manual,* Fourth Edition/Revised Printing by Keith Cunningham and Leslie Snider. Copyright © 2001 by Kendall/Hunt Publishing Company. Used with permission.

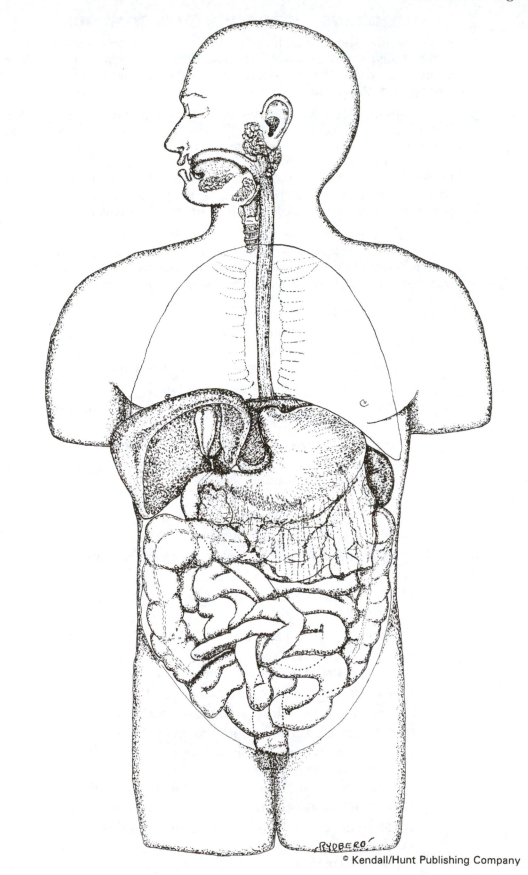

Figure 14.1 Human, Digestive System

NUTRITION AND METABOLISM QUESTIONNAIRE

It is currently estimated that a large percentage of middle-aged women and men in the United States are obese. Obesity is defined as being at least 20% over one's appropriate weight. In fact, obesity is now the number one nutritional problem among North Americans.

An alarming fact is that obesity is directly correlated with many diseases such as diabetes and atherosclerosis. Not only do we overeat, we, as a nation, invite further complications by using excessive amounts of other additives and ingredients. Salt use on average, for example, is about 20 times higher than the amount necessary to sustain our normal requirements. Sometimes we swing to the other end of the spectrum and place ourselves into similar peril by subjecting ourselves to fasting, high or low protein diets, high fiber diets, or other "extreme conditions."

Speaking of extreme conditions . . . let's take a test to see just what your knowledge is about nutrition. Don't panic . . . it won't be graded . . . this is just for your own understanding. Answer the following questions and see how well you do. You may be surprised about some common misconceptions regarding nutrition.

_____ 1. Red meat always contains more fat than poultry.

_____ 2. For complete nutrition, a balance between carbohydrates, fats, proteins, vitamins, and minerals is essential.

_____ 3. Frozen food is nutritionally inferior to fresh food.

_____ 4. Margarine is less fattening than butter.

_____ 5. Animal fats have been implicated as a possible cause of atherosclerosis and cancer.

_____ 6. Exercise reduces the risk of some types of diabetes.

_____ 7. Vegetable oils can be used safely in any amount because they contain only unsaturated fats.

_____ 8. Eggs are never a source of disease (even if eaten raw) because they are protected by their shells.

_____ 9. Fish and celery are brain food.

_____ 10. Some individuals develop allergies to certain foods.

_____ 11. Always starve a fever and feed a cold.

_____ 12. A completely vegetarian diet causes anemia, muscular weakness, and a general lack of vigor.

_____ 13. Foods grown by organic means contain more nutrients.

_____ 14. For some individuals, water is fattening.

_____ 15. Eggs, oysters, and lean (rare) meat increase sexual potency.

_____ 16. Exercise is essential for healthful weight loss.

_____ 17. Skipping a meal during the day definitely aids in losing weight.

_____ 18. Vitamins are a good substitute for many foods.

_____ 19. Raw sugar is a nutritional plus because of the additional vitamins and minerals it contains.

_____ 20. The older we get, the more calories we need for life functions.

_____ 21. Fruit juices are not fattening because they do not contain any fat.

_____ 22. Heart disease in an overweight person makes weight reduction urgent.

_____ 23. Cancer-causing chemicals (carcinogens) are produced by repeated heatings of the same cooking oil or grease.

_____ 24. It may be better for a person to remain overweight than to lose a few pounds, gain them back, and then repeat the cycle periodically.

_____ 25. Boiling fruits and vegetables in large amounts of water washes out the vitamins.

_____ 26. Dieting without medical supervision may result in sickness and endanger life.

_____ 27. An overweight person breathes 2 to 3 times more rapidly than a thin person.

_____ 28. A nourishing breakfast may help you work better in the morning, reduces nervous tension, and may prevent afternoon fatigue.

_____ 29. An individual should have three meals each day for best health.

_____ 30. Cheese and milk cause constipation.

_____ 31. Nutritional labels on packaged food are required to list the ingredients in decreasing order of abundance.

_____ 32. Some foods, such as red wine, aged cheese, nuts, and pickles can cause headaches.

_____ 33. Dieters lose weight most effectively if they limit fat intake.

_____ 34. For optimum benefit, drink eight glasses of water each day.

_____ 35. Death is simply nature's way of telling you to slow down.

Once you have completed this questionnaire refer to the Appendix which contains comments and brief explanations. It is important to recognize the many misconceptions that we have regarding the need for proper nutrition or, for that matter, just what proper nutrition involves.

From *Human Biology Laboratory Manual*, Fourth Edition/Revised Printing by Keith Cunningham and Leslie Snider. Copyright © 2001 by Kendall/Hunt Publishing Company. Used with permission.

Energy Budget

ASSIGNMENT SHEET FOR EXERCISE 15

MUST BE COMPLETED BEFORE COMING TO LAB

Next week's lab will deal with calories and energy expenditure. For the lab to be meaningful, you will need to keep a record of your physical activities and the food you ate for three days. You may use any three days of the following week; they do not need to be consecutive. The profile you will establish from this record is meant to give you a picture of your energy requirements so you should pick days that are typical of your life style at this point. For instance, if you jog or lift weights every other day, pick a day that you are active and one day that you aren't to record.

1. Keep an accurate record of your activities and food consumed on the attached sheets. For the **Activity Diary** record your activities from the time you get up through the day. Sleeping does count. Specify the activity and the length of time spent on the activity. You will calculate the energy expenditure when you come to lab since you will need to know your weight in kilograms.

2. Fill in the Dietary Chart by recording all the food and drinks along with their calories you consumed for the same three days you used for the Activity Diary. Remember to include snacks, soft drinks, and alcoholic beverages. Do not forget to add your water.

3. Please wear two piece outfits to the lab. We will be taking skin-fold measurements and will need to expose your waist and side.

4. Everyone: Bring a calculator to the lab.

Example:

Activity Chart

Activity	Total Time Spent	Factor KCAL/KG of Body Wt.	Body Wt(KG)	Total KCAL Expended
record everything for 3 days	The time you spent on the activity	p. 206 table 15.3	Body wt(KG) = BMR p. 202	This is C from p. 202 $C = B + (B \times E)$ Then multiply $C \times$ total hours to get total kcal expended
Cooking	3 hrs	0.02	59 kg	$59 + (59 \times 0.02) = 60.18 \frac{kcal}{hr}$
				so then
				$60.18 \frac{kcal}{hr} \times 3 \text{ hrs} = 180.54 \text{ kcal}$

Also see the activity chart example on p. 205.

Dietary Chart: Calories = kcal's or cal

Breakfast	Lunch	Dinner	Snacks
First Day **Date:** 1 cup OJ - 110 cal 1 cup oatmeal - 150 cal 1 cup tea - 0 cal	Salad with tomatoes, celery and carrots - 85 calories 1 T Ranch dressing - 110 cal 1 chicken breast - 165 cal 1 roll - 95 cal 1 bottle of water 500 ml - 0 cal	Shrimp boiled - 70 cal. Spinach - 240 cal. Strawberries - 30 cal. ½C Cantaloupe - 30 cal. ½ Honeydew - 60 cal. ½ Greenbeans - 15 cal. ½C water .5 L - 0 cal.	3 waters - .5 L cals Peanuts - 100 cal 24 nuts
Total Cal. for Breakfast	(+) Total Cal. for Lunch	(+) Total Cal. for Dinner	(+) Total Cal. for snacks
			= Total Cal. for 3 days

Activity Chart

Activity	Total Time Spent	Factor KCAL/KG of Body Wt.	Body Wt(KG)	Total KCAL Expended
	*			

* Must equal 24 hours

Activity Chart

Activity	Total Time Spent	Factor KCAL/KG of Body Wt.	Body Wt(KG)	Total KCAL Expended
	*			

* Must equal 24 hours

Activity Chart

Activity	Total Time Spent	Factor KCAL/KG of Body Wt.	Body Wt(KG)	Total KCAL Expended
	*			
	Total of 3 Days			
	Average			

* Must equal 24 hours

Dietary Chart: Calories = kcal's or cal			
Breakfast	**Lunch**	**Dinner**	**Snacks**
First Day Date:			
Second Day Date:			
Third Day Date:			
		Total of 3 Days Calorie Intake	
		Average of 3 Days Calorie Intake	

ENERGY BUDGET

The food you eat is digested and absorbed. The molecules made from these processes may be stored, used as structural components, or used to provide energy for your daily activities. The proportion of food energy allocated to each category is of great interest to us since it bears on our physical fitness. For example, when you eat a hamburger, the amino acids from the protein may be used to add to your muscle mass. The products from the fat may be used to build the lipid components of cells, or be broken down to make ATP, or if there is no other use for it, be converted to a storage form and packed away in adipose (fat) tissue for future use. On the other hand, if you do not eat, body cells and tissues cannot be replaced or increased and eventually you would run out of the raw materials to generate ATP.

Most of us try to balance our food intake so that we have adequate supplies for structural materials and ATP but nothing left over to store as fat. To do this, it is helpful to know what your personal energy budget is. Your energy budget is calculated by measuring the "income", calories* consumed and the "expenditures", calories used.

The body's total energy needs fall into three categories: energy to support basal metabolism, energy for muscular activity, and energy to digest and metabolize food.

Basal metabolism is the minimum amount of energy the body needs at rest in the fasting state. Certain processes necessary for the maintenance of life proceed without conscious awareness. The beating of the heart, inhaling of oxygen and the exhaling of carbon dioxide, the metabolic activities of each cell, the maintenance of body temperature, and the sending of nerve impulses from the brain to direct these automatic activities are some of the basal metabolic processes that maintain life. Their minimum energy need must be met before any calories can be used for physical activity or the digestion of food.

The **basal metabolic rate (BMR)** is the rate at which these calories are spent for these maintenance activities. The BMR varies from person to person and may change in one individual with a change in circumstance, physical condition, or age.

The second component of energy metabolism is physical activity voluntarily undertaken and achieved by the use of skeletal muscles. Contraction of muscles uses a larger number of calories, and in a moving body the heart must beat faster. This also accounts for additional caloric usage. The longer an activity lasts, the more calories are used; therefore, measurement of physical activity is expressed as calories per weight per unit time.

The final component of energy expenditure deals with processing food. When food is taken into the body, many cells become active. Muscles move the food through the intestinal tract by speeding up their rhythmic contractions, while the cells that manufacture and secret digestive juice begin to do their jobs. All these cells need extra energy to participate in the digestion, absorption, and metabolism of food. In addition, the presence of food stimulates general metabolism. All of this is referred to as **Specific Dynamic Action (SDA)** of food.

---

*The term calories in this exercise refers to food calories which are actually kilocalories or the amount of heat necessary to raise the temperature of a kilogram of water 1°C.

From *Human Biology: A Laboratory Manual*, Fifth Edition/Revised Printing by Roberta B. Williams. Copyright © 2000 by Kendall/Hunt Publishing Company. Used with permission.

To help the students out.

1. Determine your body mass (weight) in kilograms, using the following conversion factor:

    $$1 \text{ kg} = 2.2 \text{ lbs}$$
    your weight _____ kg

2. Establish your BMR by using the following approximation:

    $$1 \text{ kg uses } 1 \text{ kcal/hr}$$
    your BMR _____ kcal/hr

3. In order to establish the amount of energy expended in various activities throughout the day refer to the Energy Expenditure Factor table below. This table lists a variety of activities along with an energy expenditure factor (E) for each activity. This energy expenditure factor allows you to estimate kcal expended above and beyond your BMR while engaged in that activity.

4. For each activity you engaged in during the 24-hour time period you will determine the calories expended per hour using the following formula:

    $$C = B + (B \times E)$$

    where: C = kcal/hr expended for that activity
    B = BMR
    E = energy expenditure factor (from table)

Sample Problem:
a. mass = 150 lbs = 68 kg
b. BMR = 68 kcal/hr
c. assume this person slept 6 hours last night; the energy expenditure factor (E) for sleep is 0.1
d. the hourly energy expenditure (C) for sleeping is thus 68 + (68 × 0.1) = 75 kcal/hr
e. the total energy expended while sleeping is calculated:

$$\frac{75 \text{ kcal}}{\text{hr}} \times 6 \text{ hr sleep} = 450 \text{ kcal}$$

Physical Activity Calculations

The term physical activity is used in this exercise to mean energy expended during non-sleeping periods by skeletal muscles. The amount of energy expended will depend on the size of the body, the type of activity, and the length of the activity.

From *Cunningham & Snyder*, 1991, Kendall-Hunt.

I. PROCEDURE

Calculation of BMR

BMR is influenced by body surface area, but not weight. Two people with different shapes will have different BMR. A short, stout person will generally have a slower BMR than a tall, thin person, even if they weigh the same. The tall, thin person has a greater skin surface from which heat is lost by radiation and so must have a faster metabolism to generate the lost heat. Another factor that influences BMR is gender. Males generally have a faster metabolism rate than females, due to the greater percentage of muscle tissue in the male body. Muscle tissue is always active, while fat tissue is comparatively inactive. Age also influences BMR. The younger the person, the higher BMR. This is due to increased activity of cells undergoing division as growth occurs. After growth stops, the BMR decreases about 2 percent per decade throughout the life.

1. A scale and measuring device are set up for you to get your weight and height. The calculations you will do assume that you are fully clothed and wearing shoes with a one inch heel. For this exercise to be of value to you, be honest with your weight. You do not have to tell anyone else.

2. Use Table 15.1 to determine your body surface area. Using a ruler, draw a straight line from your height to your weight. The point where that line crosses the middle column shows your surface area in square meters. For example, for a 17-year-old male whose height is 6 feet and weight is 170 lbs., his body surface area would be 1.99 meters squared.

   Write your body surface area here _____

3. Next use the BMR constant table (Table 15.2) to find the factor for your age and sex. Multiply your surface area by this factor. Our 17-year-old male has a factor of 41.5 cals/sq. meter/hour. This multiplied by 1.99 sq. meters equals 82.6 cals/hr.

   Record your hourly BMR _____

4. Since there are 24 hours in a day, you must multiply your hourly rate by 24 to obtain the number of calories you need per day to meet your BMR requirement. The 17-year-old needs 82.6 cal/hr × 24 hrs or 1,982 calories per day to meet his BMR requirements.

   Record your daily BMR requirements _____

Table 15.1

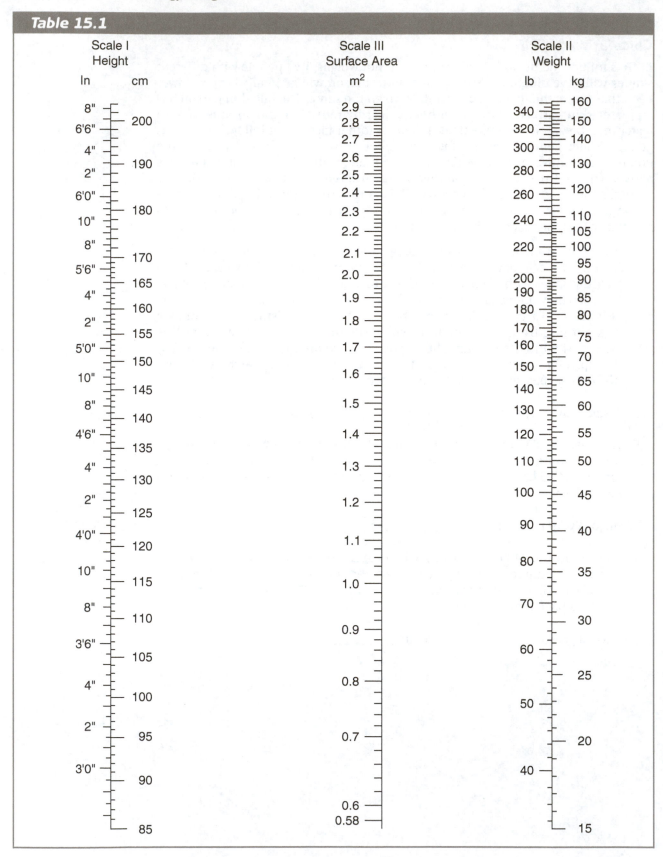

From *Human Biology: A Laboratory Manual,* Fifth Edition/Revised Printing by Roberta B. Williams. Copyright © 2000 by Kendall/Hunt Publishing Company. Used with permission.

Nomogram to estimate body surface area from height and weight. (From "Clinical Spirometry" as prepared by Boothby and Sandilford of the Mayo Clinic.) From American Physiological Society.

Table 15.2	Basal Metabolic Rate Constants					
	BMR, cal/m2/hr.				**BMR, cal/m2/hr.**	
Age	*Male*	*Female*		*Age*	*Male*	*Female*
16	42.0	37.2		32	37.2	34.9
17	41.5	36.4		33	37.1	34.9
18	40.8	35.8		34	37.0	34.9
19	40.5	35.4		35	36.9	34.8
20	39.9	35.3		36	36.8	34.7
21	39.5	35.2		37	36.7	34.6
22	39.2	35.2		38	36.7	34.5
23	39.0	35.2		39	36.6	34.4
24	38.7	35.1		40–44	36.4	34.1
25	38.4	35.1		45–49	36.2	33.8
26	38.2	35.0		50–54	35.8	33.1
27	38.0	35.0		55–59	35.1	32.8
28	37.8	35.0		60–64	34.5	32.0
29	37.7	35.0		65–69	33.5	31.6
30	37.6	35.0		70–74	32.7	31.1
31	37.4	35.0		75+	31.8	29.8

Modified from *Handbook of Biological Data* (1956)

A heavier person uses more calories performing the same tasks in the same time as a lighter person, because it takes more effort to move the additional body weight.

1. Using the activity diary you compiled last week, group each of your activities into one of the categories listed in Table 15.3. Calculate the number of **minutes** you spend on that activity.
2. Multiply the number of minutes spent on the activity by the appropriate factor for the activity and by your body weight in **kilograms** (you can determine your kilogram weight if you go back to Table 15.1) to obtain the number of calories expended for each activity. Make sure you account for every awake minute even if most of the time falls into the very light category.
3. Total all the calories expended for the three days and divide by three to obtain an average caloric expenditure per day.

If the 17-year-old's activities for the three days included:

> 1,440 minutes of sleeping (very little activity)
> 1,440 minutes of sitting (very light)
> 360 minutes of walking (light)
> 90 minutes of standing (very light)
> 450 minutes of driving (very light)
> 90 minutes of weight lifting (heavy)
> 180 minutes of jogging (moderate)

The information could be grouped as follows:

1440 min. very little activity	×	0.001 cal/kg	×	77.3 kg	= 111.3 cals
2250 min. very light	×	0.01 cal/kg	×	77.3 kg	= 1,739 cals
360 min. light	×	0.02 cal/kg	×	77.3 kg	= 557 cals
180 min. moderate	×	0.025 cal/kg	×	77.3 kg	= 348 cals
90 min. heavy	×	0.07 cal/kg	×	77.3 kg	= 487 cals
		Total expenditure for three days			= 3,242.3 cals
		Average daily expenditure			= 1,080.7 cals

Table 15.3 Daily Energy Expenditures

Type of Activity	Total Energy Expended (cal/kg of body weight/min.)
VERY LITTLE ACTIVITY: Sleep	0.001
VERY LIGHT: Sitting, standing, driving, typing, walking slowly	0.01
LIGHT: Walking at moderate speed, light housework, golf, sailing, table tennis, volleyball, carrying light loads	0.02
MODERATE: Walking fast or jogging, weeding and hoeing, carrying heavy loads, cycling at moderate speed, skiing, dancing	0.025
HEAVY: Walking quickly up hill, climbing stairs, basketball, weight lifting, swimming, football	0.07
SEVERE: Tennis, running	0.11
VERY SEVERE: Wrestling, boxing, racing	0.14

Modified from Food and Nutrition Board, Recommended Dietary Allowances, 1980.

Specific Dynamic Action Calculations

1. Add the energy for basal metabolism and that for activities and multiply the subtotal by 10% to obtain an estimated SDA. For the 17-year-old example this would be [1,982 cal (BMR) + 1,044 cal (activity)] × 0.1 = 303 cal (SDA)

Total Energy Requirement for Average Day

1. Add all three figures—BMR, activities and SDA—to obtain the number of calories needed in one day.

   BMR   _____ cals (p. 200)

   Activities _____ cals (p. 196)

   SDA   _____ cals (above)

   Total   _____ cals/day

The average 70-kilogram (155 lb.) adult male requires about 2,700 cals/day, while the average 55-kilogram (120 lb.) female needs only 2,000 cals/day. Our 17-year-old is younger and somewhat more active than the average person, and therefore has a higher energy requirement (3,329 cals/day).

II. AVERAGE DAILY CALORIC INTAKE

Now that you know how many calories you expend in an average day, you will calculate the number of calories you consume.

1. Obtain one of the calorie guide books or use the computer program to calculate the calories in the foods you listed on the dietary diary. Look up each item, estimate the amount, and record the total number of calories. Get a grand total for the three days.

2. Divide the grand total by three to obtain the average number of calories for one day.

3. Compare your daily caloric expenditure with your daily caloric intake. Measuring the amount of calories consumed is more accurate than measuring the number of calories expended. If the two figures are within 20% of each other, you are most likely maintaining your weight.

One pound of fat is equal to 3,500 calories. It takes 3,500 calories of activity to lose one pound of body weight. To lose a pound a week, you could either increase your activities so you burn off 500 calories more per day or decrease your intake by 500 calories per day. On the other hand it also takes 3,500 calories to make a pound of body weight. You could gain a pound in a week by eating one piece of pecan pie a day.

Some people seem to eat anything without gaining weight while others go through agonies to lose a couple of pounds. Caloric intake and expenditure are only part of the story here. Apparently the body has a fat "set point" it tries to maintain, much as a thermostat works to maintain room temperature. How that set point is determined, and how (and whether) it can be changed are not clear.

III. BODY COMPOSITION

Gross weight is not a good measurement of body composition. Height and weight tables are limited since what makes up weight is not taken into consideration. One person 6 foot tall and weighing 200 pounds may be overweight while another person with the same measurements may be normal. If the added pounds are fat, this is considered undesirable; however, if the weight is muscle mass, as in a body builder, the extra pounds are normal and desirable. There is a need to evaluate what is bone, muscle, and fat. Body composition refers to the lean body weight plus the fat weight. These two together make up total body weight. There are a number of ways to determine what an individual should weigh. Each method has both benefits and drawbacks.

Height and Weight Tables

Height and weight tables have been designed by life insurance companies and are constructed from data on thousands of people. The standard age used for these charts is 25 because by that age bone and muscle growth is complete. They are used because they are easy to read and people are familiar with them. The major limitation in using them is that they do not distinguish between bone, muscle, and fat. The ones that use "small frame," "medium frame," and "large frame" attempt to take into account bone size.

At best, these tables can serve as arbitrary measures for too little or too much body weight. A person more than 10 percent over the weight on the tables is overweight; a person 20 percent over is obese. Similarly, a person 10 percent below the weight on the tables is underweight.

1. Using Table 15.4 determine your desirable weight range.

Prediction Equations

Prediction equations are statistical procedures to estimate body composition from simple measurements. They are called prediction equations because they predict what someone's percent body fat is by taking measurements of the skeleton. You will use them to predict what your "ideal weight" should be based on your bone structure.

From *Human Biology: A Laboratory Manual*, Fifth Edition/Revised Printing by Roberta B. Williams. Copyright © 2000 by Kendall/Hunt Publishing Company. Used with permission.

Table 15.4   Weight and Height Tables (from the 1983 statistics of Metropolitan Life Insurance Company)

MEN

Height Feet	Inches	Small Frame	Medium Frame	Large Frame
5	2	128–134	131–141	138–150
5	3	130–136	133–143	140–153
5	4	132–138	135–145	142–156
5	5	134–140	137–148	144–160
5	6	136–142	139–151	146–164
5	7	138–145	142–154	149–168
5	8	140–148	145–157	152–172
5	9	142–151	148–160	155–176
5	10	144–154	151–163	158–180
5	11	146–157	154–166	161–184
6	0	149–160	157–170	164–188
6	1	152–164	160–174	168–192
6	2	155–168	164–178	172–197
6	3	158–172	167–182	176–202
6	4	162–176	171–187	181–207

WOMEN

Height Feet	Inches	Small Frame	Medium Frame	Large Frame
4	10	102–111	109–121	118–131
4	11	103–113	111–123	120–134
5	0	104–115	113–126	122–137
5	1	106–118	115–129	125–140
5	2	108–121	118–132	128–143
5	3	111–124	121–135	131–147
5	4	114–127	124–138	134–151
5	5	117–130	127–141	137–155
5	6	120–133	130–144	140–159
5	7	123–136	133–147	143–163
5	8	126–139	136–150	146–167
5	9	129–142	139–153	149–170
5	10	132–145	142–156	152–173
5	11	135–148	145–159	155–176
6	0	138–151	148–162	158–179

Weight at ages 25–29 based on lowest mortality. Weight in pounds according to frame (in indoor clothing weighing 5 pounds for men or 3 pounds for women), shoes with 1-inch heels.

1. Work in pairs. It is much easier to have someone measure you. Get a tape measure with centimeter markings. DO ALL YOUR MEASUREMENTS IN METRICS.
2. Measure your partner's wrists at its widest point.
3. Measure your partner's forearms, calves, and ankles at their largest circumference. Add the four measurements.

Wrist _____

Forearm _____

Ankle _____

Calves _____

Total of the 4 measurements _____

4. Divide the total by 17.07 if you are a male or 16.89 if you are a female.

$$\frac{\text{Total measurements}}{\text{17.07 or 16.89}} = \underline{\hspace{2cm}} \text{ (divided)}$$

5. Square the dividend (multiply it by itself)

(dividend) squared = _____

6. Multiply the squared dividend by your height (in centimeters, use Table 15.1 to obtain that number) and then multiply that product by 0.0111. This will give your "ideal weight" in kilograms. To convert to pounds, multiply by 2.2

(dividend) squared × height (cm) × 0.0111 = _____ kg

_____ kg by 2.2 = _____ lbs.

Skinfolds Measurements

A direct measure of the amount of body fat can be obtained by means of skinfolds test. The fat attached to the skin is roughly proportional to total body fat, and the measurements can be easily converted to percent body fat. A "target" or "ideal" weight can be calculated using the percent fat figures. The more areas of the body that are measured, the more accurate these calculations will be. In this lab we will take three measurements on women and four on men. Other measurements would require the subject to wear a bathing suit or to undress almost completely.

1. Work in pairs. Obtain a calipers designed to measure skinfolds. ALL MEASUREMENTS SHOULD BE DONE ON THE RIGHT SIDE. Firmly grasp the fold of skin between the left thumb and the other four fingers and lift. Pinch and lift the fold several times to make sure no musculature is grasped. Hold the skinfold firmly and place the contact side of the calipers below the thumb and fingers; do not let go of the fold. Take the reading to the nearest half millimeter. Release the grip on the caliper and release the fold. To make sure the reading is accurate, repeat the measurement a few times. If the second measurement is within 1–2 mm of the first, it is reliable. It takes practice to get an accurate skinfold measurement.

2. **Triceps (women only).** The triceps skinfold is measured on the back of the upper right arm, halfway between the elbow and the tip of the shoulder, while the arm is hanging loosely at the subject's side. Grasp the skinfold parallel to the long axis of the arm and lift away from the arm to make sure no muscle tissue is caught in the fold. Place the contact surface of the calipers half an inch below the fingers. Close the calipers on the skin fold. The caliper pointer will indicate the skinfold thickness in millimeters.

3. **Ilium—Hip (men and women).** Fold the skin diagonally just above the top of the hip bone on an imaginary line that would divide the body into front and back halves. Measure as described above.

4. **Abdomen (men and women).** Fold the skin vertically one inch to the right of the navel (belly button). Measure as before.

5. **Chest (men only).** Fold the skin diagonally, midway between the nipple and the armpit. Measure as before.

6. **Axilla—Side (men only).** Fold the skin vertically at the level of nipple on an imaginary line that would divide the body into front and back halves.

7. Total all your skinfold measurements.

	Women	Men
Triceps	_____	
Ilium	_____	_____
Abdomen	_____	_____
Chest		_____
Axilla		_____
Total	_____	_____

8. To estimate your percent body fat, use your skinfold measurement total and your age along with Table 15.5 (males) or 15.6 (females). This number is a "ballpark" figure because lack of expertise in using the calipers can easily produce errors.

Record % body fat _____

It has been suggested that from a health and aesthetic standpoint, adult males should have 16% or less body fat and adult females should have 23% or less body fat. At no times should the percent body fat of an adult male drop below 5% and that of an adult female below 10%. Although the news media make a point of athletes that have minimal body fat, this is not a healthy condition. Severe medical problems do occur with too little body fat as well as with too much body fat. Well conditioned athletes, such as marathon runners and swimmers usually have 10–12% body fat while football players can have as high as 19–20% body fat.

Marginal obesity would occur when an adult male's body fat increases over 20% and an adult female's over 30%.

Recent literature indicates that body weight and body build may be contributing factors to cardiovascular disease. Two new measurements, **waist-to-hip ratio** and **body mass index (BMI),** can be used to get a more complete picture of how your weight is likely to affect your health.

Table 15.5 Percent Fat Estimates for Men
Sum of Four Skinfolds, Chest, Ilium, Abdomen, Axilla

Sum of 4 Skinfolds	18 to 22	23 to 27	28 to 32	33 to 37	38 to 42	43 to 47	48 to 52	53 to 57	58 and Older
8–12	1.9	2.5	3.2	3.8	4.4	5.0	5.7	6.3	6.9
13–17	3.3	3.9	4.5	5.1	5.7	6.4	7.0	7.6	8.2
18–22	4.5	5.2	5.8	6.4	7.0	7.7	8.3	8.9	9.5
23–27	5.8	6.4	7.1	7.7	8.3	8.9	9.5	10.2	10.8
28–32	7.1	7.7	8.3	8.9	9.5	10.2	10.8	11.4	12.0
33–37	8.3	8.9	9.5	10.1	10.8	11.4	12.0	12.6	13.2
38–42	9.5	10.1	10.7	11.3	11.9	12.6	13.2	13.8	14.4
43–47	10.6	11.6	11.9	12.5	13.1	13.7	14.4	15.0	15.6
48–52	11.8	12.4	13.0	13.6	14.2	14.9	15.5	16.1	16.7
53–57	12.9	13.5	14.1	14.7	15.4	16.0	16.6	17.2	17.9
58–62	14.0	14.6	15.2	15.8	16.4	17.1	17.7	18.3	18.9
63–67	15.0	15.6	16.3	16.9	17.5	18.1	18.8	19.4	20.0
68–72	16.1	16.7	17.3	17.9	18.5	19.2	19.8	20.4	21.0
73–77	17.1	17.7	18.3	18.9	19.5	20.2	20.8	21.4	22.0
78–82	18.0	18.7	19.3	19.9	20.5	21.0	21.8	22.4	23.0
83–87	19.0	19.6	20.2	20.8	21.5	22.1	22.7	23.3	24.0
88–92	19.9	20.5	21.2	21.8	22.4	23.0	23.6	24.3	24.9
93–97	20.8	21.4	22.1	22.7	23.3	23.9	24.8	25.2	25.8
98–102	21.7	22.6	22.9	23.5	24.2	24.8	25.4	26.0	26.7
103–107	22.5	23.2	23.8	24.4	25.0	25.6	26.3	26.9	27.5
108–112	23.4	24.0	24.6	25.2	25.8	26.5	27.1	27.7	28.3
113–117	24.1	24.8	25.4	26.0	26.6	27.3	27.9	28.5	29.1
118–122	24.9	25.5	26.2	26.8	27.4	28.0	28.6	29.3	29.9
123–127	25.7	26.3	26.9	27.5	28.1	28.8	29.4	30.0	30.6
128–132	26.4	27.0	27.6	28.2	28.8	29.5	30.1	30.7	31.3
133–137	27.1	27.7	28.3	28.9	29.5	30.2	30.8	31.4	32.0
138–142	27.7	28.3	29.0	29.6	30.2	30.8	31.4	32.1	32.7
143–147	28.3	29.0	29.6	30.2	30.8	31.5	32.1	32.7	33.3
148–152	29.0	29.6	30.2	30.8	31.4	32.7	32.7	33.3	33.9
153–157	29.5	30.2	30.8	31.4	32.0	32.7	33.3	33.9	34.5
158–162	30.1	30.7	31.3	31.9	32.6	33.2	33.8	34.4	35.1
163–167	30.6	31.2	31.9	32.5	33.1	33.7	34.3	35.0	35.6
168–172	31.1	31.7	32.4	33.0	33.6	34.2	34.8	35.5	36.1
173–177	31.6	32.2	32.8	33.5	34.1	34.7	35.3	35.9	36.6
178–182	32.0	32.7	33.3	33.9	34.5	35.2	35.8	36.4	37.0
183–187	32.5	33.1	33.7	34.3	34.9	35.6	36.2	36.8	37.4
188–192	32.9	33.5	34.1	34.7	35.3	36.0	36.6	37.2	37.8
193–197	33.2	33.8	34.5	35.1	35.7	36.3	37.0	37.8	38.2
198–202	33.6	34.2	34.8	35.4	36.1	36.7	37.3	37.9	38.5
203–207	33.9	34.5	35.1	35.7	36.4	37.0	37.6	38.2	38.9

Reprinted from *The Y's Way to Physical Fitness,* 1982, the YMCA of the USA, 101 N. Wacker Dr., Chicago, IL 60606.

Table 15.6 Percent Fat Estimates for Women
Sum of Three Skinfolds, Triceps, Abdomen, and Ilium

				Age to Last Year					
Sum of 3 Skinfolds	18 to 22	23 to 27	28 to 32	33 to 37	38 to 42	43 to 47	48 to 52	53 to 57	58 and Older
8–12	8.8	9.0	9.2	9.4	9.5	9.7	9.9	10.1	10.3
13–17	10.8	10.9	11.1	11.3	11.5	11.7	11.8	12.0	12.2
18–22	12.6	12.8	13.0	13.2	13.4	13.5	13.7	13.9	14.1
23–27	14.5	14.6	14.8	15.0	15.2	15.4	15.6	15.7	15.9
28–32	16.2	16.4	16.6	16.8	17.0	17.1	17.3	17.5	17.7
33–37	17.9	18.1	18.3	18.5	18.7	18.9	19.0	19.2	19.4
38–42	19.6	19.8	20.0	20.2	20.3	20.5	20.7	20.9	21.1
43–47	21.2	21.4	21.6	21.8	21.9	22.1	22.3	22.5	22.7
48–52	22.8	22.9	23.1	23.3	23.5	23.7	23.8	24.0	24.2
53–57	24.2	24.4	24.6	24.8	25.0	25.2	25.3	25.5	25.7
58–62	25.7	25.9	26.0	26.2	26.4	26.6	26.8	27.0	27.1
63–67	27.1	27.2	27.4	27.6	27.8	28.0	28.2	28.3	28.5
68–72	28.4	28.6	28.7	28.9	29.1	29.3	29.5	29.7	29.8
73–77	29.6	29.8	30.0	30.2	30.4	30.6	30.7	30.9	31.1
78–82	30.9	31.0	31.2	31.4	31.6	31.8	31.9	32.1	32.3
83–87	32.0	32.2	32.4	32.6	32.7	32.9	33.1	33.3	33.5
88–92	33.1	33.3	33.5	33.7	33.8	34.0	34.2	34.4	34.6
93–97	34.1	34.3	34.5	34.7	34.9	35.1	35.2	35.4	35.6
98–102	35.1	35.3	35.5	35.7	35.9	36.0	36.2	36.4	36.6
103–107	36.1	36.2	36.4	36.6	36.8	37.0	37.2	37.3	37.5
108–112	36.9	37.1	37.3	37.5	37.7	37.9	38.0	38.2	38.4
113–117	37.8	37.9	38.1	38.3	39.2	39.4	39.6	39.8	40.0
118–122	38.5	38.7	38.9	39.1	39.4	39.6	39.8	40.0	40.5
123–127	39.2	39.4	39.6	39.8	40.0	40.1	40.3	40.5	40.7
128–132	39.9	40.1	40.2	40.4	40.6	40.8	41.0	41.2	41.3
133–137	40.5	40.7	40.8	41.0	41.2	41.4	41.6	41.7	41.9
138–142	41.0	41.2	41.4	41.6	41.7	41.9	42.1	42.3	42.5
143–147	41.5	41.7	41.9	42.0	42.2	42.4	42.6	42.8	43.0
148–152	41.9	42.1	42.3	42.8	42.6	42.8	43.0	43.2	43.4
153–157	42.3	42.5	42.6	42.8	43.0	43.2	43.4	43.6	43.7
158–162	42.6	42.8	43.0	43.1	43.3	43.5	43.7	43.9	44.1
163–167	42.9	43.0	43.2	43.4	43.6	43.8	44.0	44.1	44.3
168–172	43.1	43.2	43.4	43.6	43.8	44.0	44.2	44.3	44.5
173–177	43.2	43.4	43.6	43.8	43.9	44.1	44.3	44.5	44.7
178–182	43.3	43.5	43.7	43.8	44.0	44.2	44.4	44.6	44.8

Calculating Your BMI

Multiply your weight in pounds by 700, divide by your height in inches, and then divide by your height again.

Weight (lbs) × 700/Height (in.)/Height (in.) = BMI
How the BMI value is related to cardiovascular disease
BMI of 20 or less—very low to low risk
BMI of 25 to 30—low to moderate risk
BMI of 30 or more—moderate to high risk

Calculating Your Waist-to-hip Ratio

Use the tape measure to find the circumference of your waist (in inches) at its narrowest point when your stomach is relaxed. Then measure the circumference of your hips (in inches) at their widest (where your buttocks protrude the most).

Waist _____ in.   Hips _____ in.

Divide your waist measurement by your hip measurement.

Waist/hip = _____ Waist-to-hip ratio

Women that have a waist-to-hip ratio of less than 0.8 are at low risk of cardiovascular disease.

Men that have a waist-to-hip ratio of less than 0.95 are at low risk of cardiovascular disease.

BMI and waist-to-hip ratios determine your likely range of risk. But where you fall within that range depends on other factors such as age, blood pressure, blood cholesterol levels, smoking, and the amount of exercise you get. Bear in mind that these factors give you only an approximation of your risk. Genetics also plays a role.

From *Human Biology: A Laboratory Manual*, Fifth Edition/Revised Printing by Roberta B. Williams. Copyright © 2000 by Kendall/Hunt Publishing Company. Used with permission.

REVIEW QUESTIONS

1. What three factors influence your BMR and how did you take all three of these factors into account when you calculated your BMR?

2. Calories expended for physical activities _____

   Calories expended for BMR _____

   Did you expend more calories for your physical activities or for your BMR? Do you think this situation is the same for most people? Why or why not?

3. If you jogged for 20 minutes or walked moderately for 40 minutes would you expend the same number of calories? Why or why not?

4. What one item from your dietary diary contributed the most calories to your diet? How many calories?

5. Average number of calories you ingested per day _____

   Average number of calories you expended per day _____

   How do the calories you ingested balance with the calories your expended?

6. Ideal weight according to prediction equation _____

    Weight range from height/weight tables _____

    Do you think the weight and height tables give *you* a more accurate picture of your "ideal weight" than the prediction equation?

7. Because of the lack of experience of the measurer, you most likely have not gotten an accurate percent body fat measurement. Have your lab instructor look at your calculated percentage and see if he/she thinks it is close to expected. If it isn't, they will remeasure you.

8. Often during an exercise program individuals do not lose weight but look better physically. If body fat measurements were taken at the beginning and end of the program, most likely, this person would have decreased their percent body fat. Why didn't they lose weight?

9. Did your BMI measurement and your waist-to-hip ratio put you in the same category of risk of cardiovascular disease?

10. What do you think was the most valuable part of this exercise for you personally? Why?

Nervous System

The human nervous system is divided into two basic parts: (1) *The Central Nervous System* (CNS), which consists of the brain and spinal cord, and (2) *The Peripheral Nervous System* (PNS), which consists of all the nervous tissue outside of the CNS.

I. BASIC CONCEPTS AND DEFINITIONS OF THE NERVOUS SYSTEM

1. **Neuron** is the nerve **cell.** It is the basic functional unit of the nervous system. Neurons have numerous cytoplasmic processes called **axons** and **dendrites.** Neurons are named for the number of processes that they have, i.e., unipolar, bipolar, and multipolar. They can be named for their function, i.e., afferent (sensory), and efferent (motor).

2. A **nerve** is a bundle of nerve fibers (autonomic, sensory, and motor) which course together outside of the CNS. A single nerve **fiber** is a long cellular process: an **axon** of a motor neuron or **peripheral process** of a sensory neuron. An axon takes an impulse away from a nerve cell body to a muscle or other nerve, while a dendrite carries the impulse into the nerve.

3. **Cell body** is the part of the neuron that contains the cell nucleus and from which the processes extend.

4. **Ganglia** are **groups** of cell bodies outside of the CNS, e.g., dorsal root ganglion.

5. **Nerve tracts** are groups of axons coursing together inside of the CNS. They tend to have common origins, functions, and terminations. They either ascend or descend in the spinal cord and brain stem.

6. **Grey matter** consists of unmyelinated nerve fibers and cell bodies.

7. **White matter** consists of myelinated nerve fibers.

8. **Afferent** means an impulse or nerve fiber going toward the CNS, usually from a receptor. Another meaning for afferent in the nervous system is **sensory.**

9. **Efferent** means an impulse or nerve fiber going away from the CNS to an effector organ (muscle or gland). Another name for efferent in the nervous system is **motor.**

10. **Synapse** is a junction between a neuron and effector (another neuron or a muscle cell). Presynaptic refers to the incoming neuron, postsynaptic refers to the neuron beyond the synapse.

11. **Autonomic NS** consists of those motor fibers that pass in nerves supplying the organs, glands, and smooth muscle of the body. This system has two divisions that oppose each other's action: **sympathetic** and **parasympathetic.** It also includes collections of cell bodies called autonomic ganglia.

12. **Cerebral cortex** is the outer layer of grey matter in the cerebrum.

13. **Nuclei** are functionally related groups of cell bodies found in the CNS.
14. **Mixed nerve** is the type of nerve that has both sensory (afferent) and motor (efferent) fibers. Most of the nerves in the body are like this. These also may contain autonomic nerve fibers.
15. **Funiculus** is a region of white matter in the spinal cord that contains nerve tracts.
16. **Fasciculus** is a subdivision of a funiculus (a nerve tract).
17. **Cranial nerves** consist of twelve paired nerves that come off the cerebrum, diencephalon, and brainstem.
18. **Spinal nerves** consist of 31 paired nerves that come off the spinal cord and pass into the periphery to become peripheral nerves.
19. **Plexus** consists of a network of nerves formed by the ventral rami of spinal nerves in the cervical, brachial, and lumbo-sacral regions of the cord. Spinal nerves exchange fibers in plexuses. (Autonomic plexuses also exist.)
20. **Effector** is a muscle or gland that produces a response to an efferent impulse.
21. **Receptors** are the general specialized endings of the peripheral processes (fibers) of sensory neurons. They are found both in the periphery (skin, muscles) and in the central areas (membranes, organs, tubes, etc.). They respond to stimuli.
22. **Basal nuclei** (basal ganglia) are nuclei of gray matter between the diencephalon and the white matter of the cerebrum. They are the **caudate nucleus,** the **lentiform nucleus** and a few additional nuclei we won't name in lab. The exact functions of these bodies are unknown but it is agreed that caudate and lentiform nuclei help **stop** and **start** voluntary muscle movements that are dictated by other parts of the brain.
23. **Corpus striatum** is comprised of the caudate nucleus and the lentiform nuclei.
24. **Ventricles** are the four, continuous hollow spaces in the interior of the brain. They are filled with cerebrospinal fluid (CSF).
25. **Pyramidal tracts** consist of descending fibers from the primary motor area of the cerebral cortex that pass through the medullary pyramids on their way to ventral horn cells in the spinal cord. They supply large muscle groups and control voluntary movements.
26. **Somatic nerve fibers** are both sensory and motor (afferent and efferent) fibers, which pass from and to the skin and skeletal muscles.
27. **Visceral nerve fibers** are both sensory and motor, supplying the smooth muscle in blood vessels and tubes, glands, and organs of the body. They also innervate cardiac muscle in the heart. The visceral fibers that are efferent to these structures are the fibers of the autonomic nervous system. Visceral afferent fibers pass from organs, glands, and smooth muscle into the central nervous system.
28. **Choroid plexuses** are a combination of ependymal cells and blood vessels from pia mater. Cerebrospinal fluid is formed (from the blood) in these structures. Each of the four ventricles contains a choroid plexus.
29. **Meninges** are the three-layered coverings surrounding the entire CNS.

II. INTRODUCTION

The nervous system is among the most complex systems in your body. It is much more complex than the coverage in this exercise, and you should be aware that the material presented is highly simplified. The chief functions of the nervous system are the integration of activities of several parts of the body and the conduction of impulses. The first of these functions actually includes a multitude of activities dealing with not only intelligence and memory, but also perception, will, appetite, body temperature, water metabolism, coordination, motor activity, and the control of heartbeat, breathing digestion, and blood pressure, to name only a few.

The nervous system consists of the **brain** and **cranial nerves**, the **spinal cord** and **spinal nerves**, and certain outlying **ganglia** and **fibers**. The nervous system connects **receptors** (the sense organs) with **effectors** (muscles and glands) and conducts impulses from one to the other. The coordination of these various activities is truly a phenomenal feat!

In this laboratory exercise, you will examine sheep brains and slides of the spinal cord, sensory neurons, and motor neurons. Begin by obtaining a sheep brain from your instructor.

III. THE BRAIN

The brain represents the enlarged, anterior end of the spinal cord. The detailed anatomy of the brain is exceedingly complex, and we shall observe only seven main regions. Examine a sheep brain that has been sectioned along the midsagittal line, compare it with Figure 16.1, and locate the following parts:

Cerebrum—This is the largest component of the mammalian brain; it is divided into the right and left cerebral **hemispheres** by the **longitudinal fissure.** The outer portion of the cerebrum is the **cortex**, which is comprised of **gray matter.** Sensory perception, will, memory, and intelligence are centered in this portion of the brain. Dorsally and laterally, note the long groove that appears to separate the cerebral hemispheres into anterior and posterior sections. This is the **fissure of Rolando.** Posterolaterally to the fissure is the **temporal lobe** where the sense of hearing is centered. There are three other important lobes. These are the **frontal lobes**, centers for association; **parietal lobes**, centers for sensory and motor functions of the legs, trunk, arms, and head; and the **occipital lobes**, centers for vision and speech.

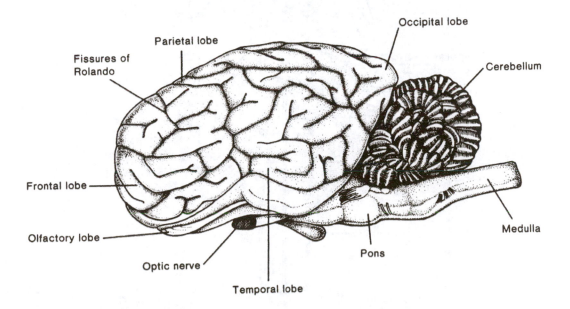

Figure 16.1a   Sheep brain, lateral view.

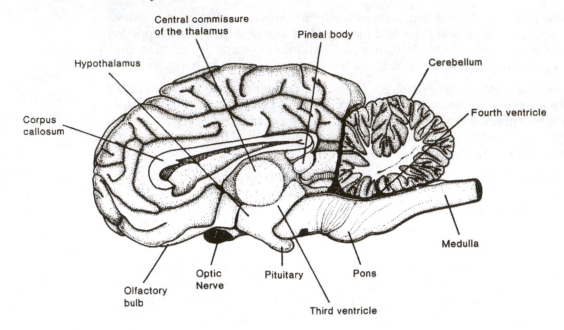

Figure 16.1b   Sheep brain, longitudinal section.

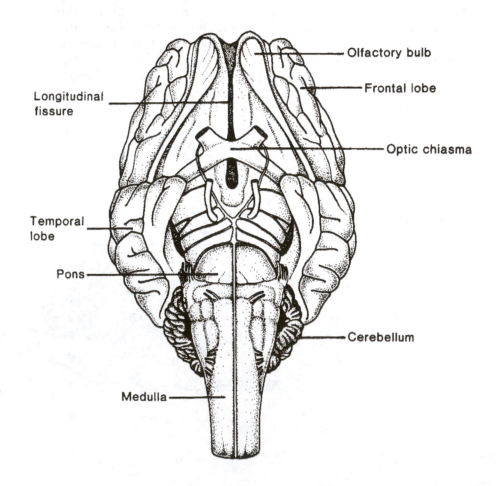

Figure 16.1c   Sheep brain, ventral view.

Olfactory lobes—At the anteroventral end of the brain, you will note two elongated tracks known as the olfactory lobes. These are the centers for the sense of smell. They are much reduced in the human brain, and this should lead you to the conclusion the sheep smell better than humans.

Thalamus—The thalamus consists of two lobes that are connected by the **central commissure.** The thalamus serves as a "relay station" for impulses carried between the brain centers and the cerebellum that you will examine later. It is a center for the sensations of pain or pleasure.

Hypothalamus—The hypothalamus forms the floor of the brain and connects to the **pituitary gland,** a small gland extending from the floor of the brain. The hypothalamus contains controlling centers for appetite, body temperature, water metabolism, pleasure, reproductive behavior, hostility, pain, and blood pressure. Therefore, it is often regarded as having a primary function in maintaining homeostasis.

Pons—The pons is a bulb-like enlargement on the brain stem. It serves as a passageway for nerve fibers connecting the medulla and the higher brain regions.

Cerebellum—This region of the brain is located posterior to the cerebrum. The cerebellum functions below the level of consciousness on coordinating equilibrium and motor activity.

Medulla oblongata—This is the most posterior part of the brain and connects the higher brain regions with the spinal cord. It is the major center for the autonomic nervous system. The latter controls "involuntary actions" such as heart rate and breathing. Generally, this involves two nerves, one which speeds up the activity and another which slows activity.

Ventricles—There are four ventricles or cavities within the brain. The first and second ventricles are located one in each cerebral hemisphere. The third ventricle is a narrow slit beneath the first two and between the thalamic lobes. The fourth ventricle lies between the cerebellum and the pons and medulla.

IV. SPINAL CORD

You will examine microscopic preparations of the spinal cord. The cord is comprised of gray matter and white matter. The gray matter is easily recognized because it picks up more dye in the staining process. Notice that the gray matter forms the shape of a butterfly. Two sets of extensions emanate from the spinal cord. These are the dorsal horns and the ventral horns. Groups of cells connected with the dorsal horns may be present in the section. These are the dorsal root ganglia. The ventral horns contain motor neurons. You may be able to identify the cell body, axon, and dendrites of these. Note that the spinal cord is hollow. This lumen contains **cerebrospinal fluid.**

V. NEURONS

Examine slides of neurons. Note the cell body which contains the nucleus. You should also be able to observe the short dendrite leading to the nerve body and the longer axon leading away. The impulse is transmitted along the fiber from dendrite to axon to the dendrite of another nerve, and so on. Examine Figure 16.2 of sensory, inter, and motor neurons.

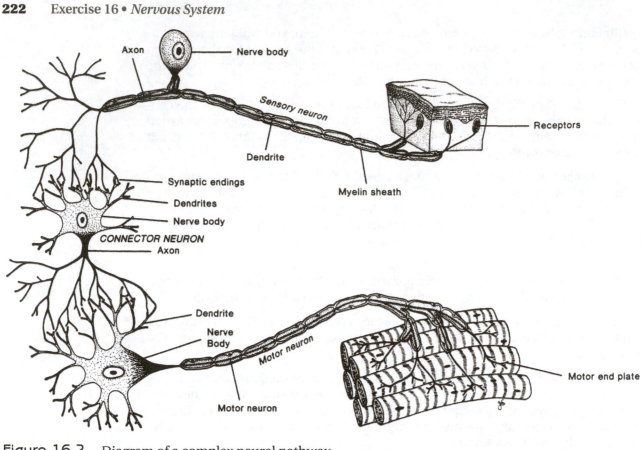

Figure 16.2   Diagram of a complex neural pathway.

REVIEW QUESTIONS

Provide the portion of the brain where each of the following is centered:

1. reproductive   _____

2. autonomic n.s.   _____

3. sense of smell   _____

4. vision   _____

5. water metabolism   _____

6. sense of hearing   _____

7. equilibrium   _____

8. center for association   _____

9. intelligence   _____

10. heart rate   _____

11. Information (nerve transmission) may pass from the CNS to the PNS.

    _____ (Yes or No)

12. Nerve cells transmit impulses from one cell to another. _____ (True or False)

13. All reflex activity must involve the spinal cord. _____ (True or False)

14. There are sensory receptors in the skin and eye. _____ (True or False)

15. There are skin receptors for pain, touch, and temperature. _____ (True or False)

16. What is the longest cell in the body? _____

17. The reflex mechanism allows for muscle contraction before the brain

    receives the touch impulse. _____ (True or False)

18. Briefly explain the all or none law.

19. Nerves that have lipid/complex coverings transmit slower than those lacking

    the lipid/complex. _____ (True or False)

20. The space between two nerve cells is approximately _____.

---

21. Impulses pass from one nerve cell to another because a chemical bridges the

    space. _____ (True or False)

22. Descending tracts carry information from a higher center down the spinal

    cord. _____ (True or False)

23. _____ carry information from one cell to dendrites of another cell.

24. The junction that occurs in #8 is termed a _____.

25. Memory gives us a record of our past. _____ (True or False)

26. Each hemisphere of the brain is associated with sight and movement on the

    opposite side of the brain. _____ (True or False)

27. In the human the _____ side of the brain contains two separate speech
    centers.

28. In the human the _____ side (half) of the brain handles spatial and
    visual duties.

29. The right side of our brain determines our muscle ability. _____ (True or
    False)

Label the parts of the brain. Include the lobes of the cerebrum, cerebellum, the parts of the brain stem, the corpus collosum, the ventricles, and the pituitary gland.

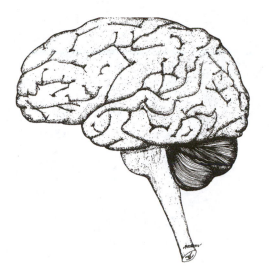

Figure A.   Lateral View

Label all the interior portions of the brain.

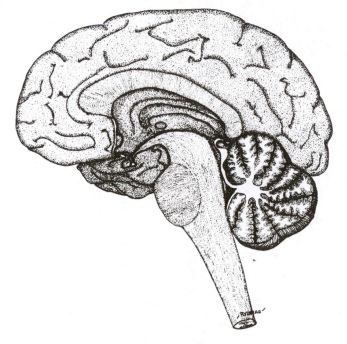

Figure B.   Sagittal View

Label all the parts of the neuron including the nerve cell body, nucleus, cytoplasm, dendrites, axon, Schwann cells, nodes of Ranvier, and myelin sheath.

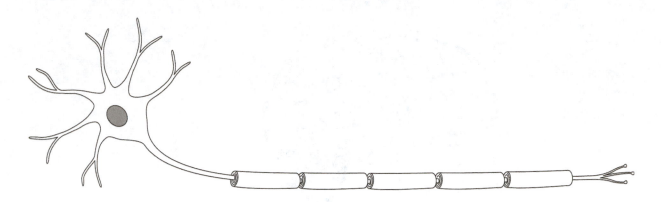

Nervous System Physiology

I. INTRODUCTION

What is red? What I'm after is not a list of all red objects, but rather what it means to be red. I've asked this question to hundreds of students, and I usually get no immediate verbal response. Instead, they stare at me with one of two looks on their faces: confusion that says, "my professor is an idiot who doesn't know what red is;" or, from those who know me a little, an amused, pitying grin anticipating a stupid joke. But when I press them and they accept that I'm after a real response, I get something like, "red is a color objects can have," or from those who have had at least one science course, "red objects reflect red light and absorb all other colors."

More interesting than these answers is the tone of voice typically used. It's almost always confused and hesitant, which I tend to interpret as meaning the student can't figure out why a professor doesn't know such a simple thing. However, students often tell me later that they were hesitant because, as one student told me, "I *knew* it couldn't be that easy."

That's an insightful position because the typical answers are not right. Consider the first one, red is a color *objects* can have—implying red is a property of the object. It isn't. The second answer makes red into a property of light. It's not that, either. Rather, red is a property of the human *brain*. Red is a tag the brain puts onto impulses from a certain type of cell in the retina. These cells send impulses when they detect light with a wavelength around 680 nm (nanometers). Therefore, colors are made by the brain so that it can comprehend different wavelengths of light intuitively.

That understanding motivates the main point of today's lab. Everything about the world, everything you think you know, is a fabrication. It's a darn good one in most cases, but still a fabrication. That's because the world does not interact directly with your brain. Rather, the brain depends on sense organs to send it the data it needs to construct a model of reality, and it's that *model* we use to decide how to act. If for some reason the model is flawed, either because of a malfunctioning sense organ or damaged brain, a person reacts to the world incorrectly, which threatens that person's health directly.

Consider a drunken man as he staggers about a restaurant looking for the bathroom. The alcohol he drank is causing a portion of his brain called the cerebellum to malfunction. One of the cerebellum's main functions is to coordinate body movements. It gathers data from nerves in muscles and joints that tell it the angle of every joint. It uses these data to build a model of the body's position in space. In the drunk, however, some of these impulses come late or not at all, and some that do make it are not processed correctly, all a direct result of the nerve toxin called alcohol. Because of flaws introduced into the drunk's brain's model of his own body, his muscles don't work in a coordinated fashion and he staggers about. He can't find the bathroom because of other mistakes caused by malfunctioning sensory and data processing organs.

Clearly sensation is absolutely central to psychology and a properly functioning nervous system in general. But, that's only part of the story. In addition to perceiving the world you also have to interact with it. Some of these interactions are extremely complex and require cognition. However, others are very simple and automatic. These we call *reflexes*. Today's lab is an introduction to sensation and simple reflexive responses.

II. SENSATION

In the following exercises perform the task and then answer the questions before moving on. However, you don't have to answer all the questions during lab. (If you do you might run out of time.) Instead, view the questions as exercises. To answer them you should consult your textbook and other resources in the laboratory, library, and internet as references.

Vision

A. **Blind Spot:** Hold this paper about 15 inches from your face, close your left eye and focus your right eye on the cross below (figure 17.1). Slowly move the page closer to your eye. At some point the dot will disappear. When that happens, have your partner measure the distance from your eye to the page. Repeat the procedure with your other eye, this time focusing on the **dot** and measuring when the **cross** disappears. Record your results in the space below.

Figure 17.1   Blind spot test.

Distance where image disappears:

Left eye _____ Right eye _____

(a) What part of the eye *specifically* detects light?

(b) Investigate, using your textbook and other resources, the cause of the blind spot. Place your answer in the space below. (NOTE: You won't necessarily find an explanation for this phenomenon in any book. You'll have to deduce it from an understanding of how the eye works. Look for a portion of the light-detecting tissue that cannot actually detect light.)

B. **Near-point accommodation:** Obtain a penny from the supply area and hold it at arm's length from your eyes with your right hand. Close your left eye and look at the markings on the penny with your right. Bring the penny towards your right eye. When the markings start to go blurry, stop and have your partner measure the distance from eye to penny. This distance is your right eye's *near point.* Next, measure the near point of your left eye and record your data below:

Left eye _____ Right eye _____

(a) The majority of the focusing of light is done by what portion of the eye? (HINT: it's not the lens. Consult the resources you have available.)

(b) How does the shape of the lens change when a person looks from a distant object to a nearby one? (Again, use the resources available.)

(c) Table 17.1 gives average near points for people of various ages. What happens to near point as a person ages? Why does this occur?

C. **Astigmatism:** Obtain an astigmatism chart from the supply area. Remove your glasses if you have them and look at the chart from the specified distance and record below which lines, if any, appear lighter in color or blurred. Replace your glasses if you have them and look again. Are these lines still blurry? If you have contact lenses instead of glasses do not remove them for this test.

Table 17.1 Mean near points for different ages.		
Age	*in*	*cm*
10	3.0	7.5
20	3.5	8.8
30	4.5	11.3
40	6.8	17.0
50	20.7	51.8
60	33.0	82.5

(a) Which lines, if any, appeared blurry or lighter?

Write "none" if all lines are equally focused _____

(b) How do near- and far-sighted eyes differ in shape?

(c) What is astigmatism?

D. **Afterimages:** Obtain triangles of red, green, and blue colored paper and a large white sheet of paper. Take these materials outside in the sun or to another bright location. Place a red triangle on the white paper and stare at a point of the triangle for at least one minute. Do not move your focus from place to place, and try to minimize blinking. Then, quickly shift your gaze to a blank part of the white paper and blink frequently. You should see a ghostly afterimage of the triangle or part of it. Record the color of the afterimage in the space below and repeat this procedure with the two remaining colors.

Afterimage of red _____

Afterimage of blue _____

Afterimage of green _____

(a) How do rod and cone cells of the retina differ in shape and function?

(b) What colors do cone cells respond to directly?

(c) Yellow is not a color that cones respond to directly. Then, how does the eye perceive yellow?

(d) Suppose you wanted to design a picture such that the afterimage is the American flag. What colors would you have to make the blue field and red stripes?

Blue field _____

Red stripes _____

III. HEARING

Unfortunately most hearing tests need a quiet environment that cannot be achieved in most laboratories. However, a test called the **Weber test of hearing loss** actually requires a noisy environment.

1. Obtain a low C tuning fork from the supply area.
2. Gently but firmly strike the tuning fork on the palm of your hand. If you strike too hard you will hear high-pitched harmonics that will invalidate the test. You should not be able to hear the tuning fork unless it is very close to an ear, and then it should be about as low a pitch as anything you can hear.
3. Hold the handle of the vibrating tuning fork firmly against the bony upper portion of the bridge of your nose. Close your eyes and visualize the direction the sound seems to be coming from.
4. A person with no hearing loss will localize the sound as coming from directly in front of the face. However, to a person with hearing loss, the sound will be slightly lateralized right or left. The worse the hearing loss the worse the deflection.
5. Repeat the test with a wad of cotton in one ear. Describe how the cotton affects the results of the test.

6. Name the three bones of the middle ear and describe their function.

7. What portion of the ear actually converts sound waves into nervous impulses?

IV. TOUCH

Touch receptors are specialized nerve endings in the skin. The more of these nerve endings there are per unit skin area, the more sensitive that portion of the skin is.

Table 17.2 Touch receptor density results.

Body region	Min. Dist.
Forearm	
Back of neck	
Fingertip	
Upper arm	
Palm	

The following procedures will allow you to compare touch receptor density of various parts of the body surface.

1. Have one partner sit with his or her eyes closed. The other partner will probe the first's skin in various locations with calipers to determine *the shortest distance between the caliper points that the skin can still detect as two distinct points.*
2. Starting with the palm-side of the forearm, partner 2 will touch partner 1's skin with the calipers. Partner 1's eyes are closed all the while. Partner 2 will vary the distance between caliper tips. Each time partner 1 is touched, he or she will indicate whether they feel one or two points of pressure. With this method, determine the smallest distance between tips that can still be felt as two points.
3. To keep partner 1 honest, partner 2 should sometimes touch the skin with one caliper tip, other times with both.
4. Repeat this procedure on the following skin locations: the back of the neck; fingertip; upper arm; and palm of hand. Fill in your data on Table 17.2.
5. What exactly do the following types of cells sense?

Meissner's corpuscles   _____

Ruffini's corpuscles   _____

Krause's end bulb   _____

Pacinian corpuscles   _____

Bare nerve endings   _____

V. REFLEXES

Biologists have come to recognize certain patterns of phenomena within the nervous system that help us organize our knowledge. One of these patterns we call a *reflex arc.* Reflex arcs are coordinated responses to stimuli external to the body. The body can only respond to stimuli it is aware of. That awareness requires the stimulus to interact with a body part called the *sensor.* The sensor, which is always a sense organ of some sort, responds to the stimulus by emitting nervous or chemical signals. These signals travel to another body part called the *integrator,* often

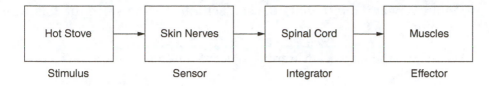

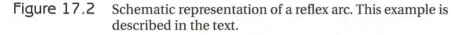

Figure 17.2   Schematic representation of a reflex arc. This example is described in the text.

part of the central nervous system but not always, that measures the level of stimulus and coordinates the proper response. After "deciding" what the proper response is, the integrator then signals a third body part, called the *effector* that performs the response (figure 17.2).

The classic example of a reflex arc occurs when a person touches a hot stove (figure 17.2). Bare nerve endings in the skin sense the temperature and send that information to the spinal cord. The spinal cord measures the intensity of the signal, which is proportional to the temperature. If the signal is above some threshold, the spinal cord institutes "emergency measures" and without consulting the brain sends a volley of signals to muscles in the arms and shoulders. The muscles contract in response, pulling the hand away from the offending stove. In this example, bare nerve endings are the sensor, the spinal cord is the integrator and the muscles are effectors.

A. Patellar ligament reflex

1. The *patellar ligament* stretches just below the knee and connects the *patella* (knee cap) to the upper portion of the *tibia* (shin bone).
2. Have one partner sit on the lab bench with his or her legs dangling free. With a reflex hammer, partner 2 should gently but firmly tap partner 1's right patellar ligament. Then repeat with the left leg. What was the response? Was it immediate or delayed?

3. Repeat these procedures but this time have partner 1 close his or her eyes, clasp his or her hands together and try to pull them apart. Without warning tap partner 1's patellar ligament on a leg chosen at random. Did this response differ from the previous one? If so, how?

4. Repeat again, this time with the subject concentrating on the reflex itself and visualizing his or her leg moving outward as the test is done. Start with the right leg and give the subject warning as you tap the ligament. Record any differences in this response compared to the previous two below.

5. Finally repeat the test while the subject attempts to solve the following equation:

$$2 \times 5 \times 6 \times 21 \times 10 \times 3 \times 322 =$$

Do not give the subject any warning. Record your observations below.

6. Using your textbook and resources in the lab as references, determine what the sensor is in this reflex.

7. What is the effector(s)?

B. Doorjamb reflex

1. Stand in an open doorway with your arms at your sides. Then push your arms out so that the backs of your hands are on the doorframe.
2. Press outward on the doorframe as if you were trying to raise your arms laterally over your head.
3. Push in this manner against the doorframe for one full minute.
4. Relax and step forward out of the doorway. How do your arms feel? Record your response below.

5. What is the sensor in this reflex?

6. What is the effector?

C. Pupillary constriction

1. Align a lamp that is switched off to shine into one partner's eye.
2. Turn the lamp on. What happened to the size of the pupil?

3. Did the pupil of the other eye, the one without the bright light shining in it, respond similarly or differently?

4. What is the sensor in this reflex?

5. What is the effector?

D. Accommodation reflex

1. Have one partner focus on an object as far away as possible. The other partner should be watching partner 1's pupil diameter.
2. Observe how partner 1's pupil diameter changes when they switch their focus to an object, a pen or pencil tip, held about 5 inches from the face. Record your observations in the space below.

3. What is the sensor in this reflex?

4. What is the effector?

E. Nystagmus reflex

1. Partner 1 should sit in a swivel chair well away from the lab benches or any other object in the lab. This partner should then close his or her eyes and tilt their head forward slightly.
2. Partner 2 will then spin partner 1 rapidly *but carefully* for 20 to 30 seconds. If the person being spun becomes uncomfortable, stop immediately.
3. When the spinning stops, look in partner 1's eyes and describe below what you see happening.

4. What is the sensor in this reflex? (HINT: it has nothing to do with the eye, but rather the ear).

5. What is the effector?

VI. TASTE

For fun have your lab partner hold their nose or put on nose clips and close their eyes. Place a piece of food on their tongue. They should roll the food in their mouth with their tongue and chew on it with their teeth. They should try to identify the food without smelling or seeing it. Foods with similar textures should be used. Prunes and raisins or apples and radishes or egg white and avocado can be tested.

Label the parts of the eye and ear: Cornea, Retina, Pupil, Lens, Iris, Aqueous humor, Fovea centralis, Blind spot/optic disk, Optic nerve, Vitreous humor, External auditory canal, Tympanic membrane, Cochlea, Stapes, Malleus, Incus, Eustachian tube, Semicircular canals

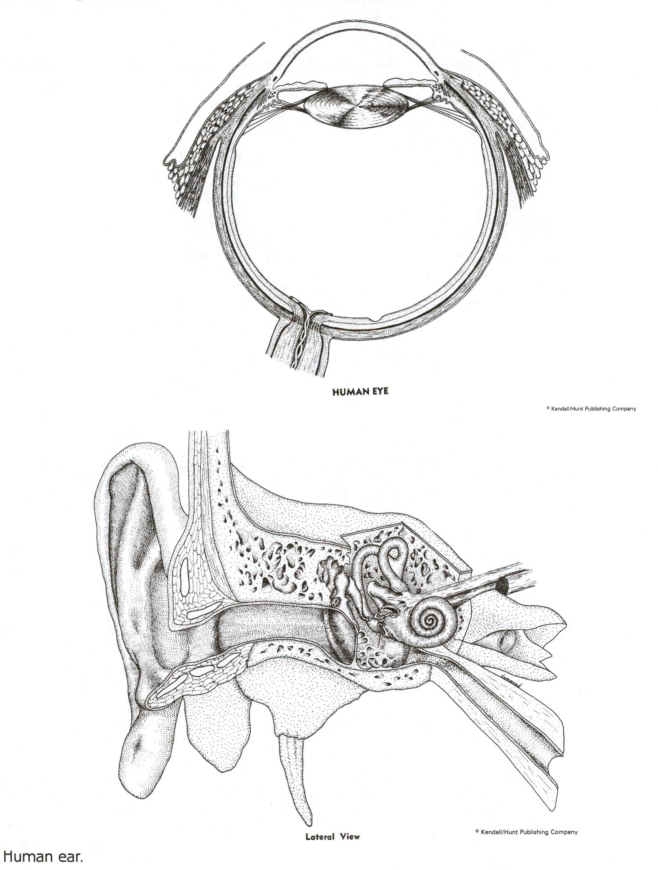

HUMAN EYE

© Kendall/Hunt Publishing Company

Lateral View   © Kendall/Hunt Publishing Company

Human ear.

Endocrine System

I. INTRODUCTION

The function of the endocrine glands is to integrate, correlate, and control body processes (metabolism and homeostasis) by **chemical means via circulation**.

An endocrine gland has morphologically distinct cells that produce specific chemicals at the ribosomal area, and these chemicals exert a specific effect at some cellular site in the body, called the **target organ**.

A hormone is a product of an endocrine gland and is active at some receptor site of a given tissue.

In man and other vertebrates many different types of molecules have hormonal activity and new hormonally active compounds are continually being discovered. In addition to the traditional hormones (from the endocrine glands) there are other regulatory substances that can influence the behavior of body cells. Despite this diversity, certain generalizations can be made about the chemical basis of hormonal action and structure.

Scientific and medical researchers first became interested in hormones because of selected diseases associated with malfunctions of the various endocrine glands. It has become evident that hormones play an important part in the functions of virtually every living system.

There are three types of hormones: **amines**, **proteins** and **steroids**. A typical protein hormone that is familiar to everyone is insulin, which is synthesized by specific pancreatic cells and consists of 51 amino acid units. The protein hormones are comparatively large molecules (**macro**), typically with a molecular weight of about 6,000. Because of their macromolecular size they **do not** readily pass through the cell membrane and in fact they do not enter the cell. Instead they act at the cell membrane surface. The macromolecular protein can influence a great variety of intracellular processes, but only indirectly. The steroids and amines, typically with a molecular weight of about 500, can **easily** pass through the cell membrane and thus *directly* effect the metabolism of the respective cell.

Biochemists have recently observed a compound (adenosine 3'5' monophosphate or **cyclic AMP; cAMP**) that has only regulatory functions. cAMP is synthesized from ATP in a reaction catalyzed by the enzyme adenyl cyclase. This cyclic nucleotide is degraded in a reaction catalyzed by a phosphodiesterase enzyme.

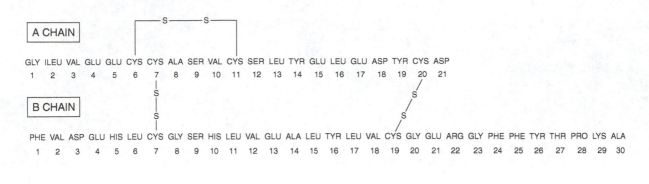

Pancreatic hormones

II. PROTEIN HORMONES

In the cells of vertebrates, adenyl cyclase is found as a component of the **cell (plasma) membrane**. In this location, **cAMP production can be stimulated by various external influences**. In fact, cAMP was first detected during studies of the mechanism by which the hormone epinephrine promotes glycogen breakdown in the liver. It was found that epinephrine acts by enhancing cAMP production at the plasma membrane of the liver cells. The cAMP functions within the cell as a **second messenger**, triggering the actual metabolic changes associated with the presence of the epinephrine at the receptor site on the cell membrane surface.

As animal research continued we have been able to observe that many vertebrate hormones function simply by stimulating cAMP production in specific target cells. The specificity of response to hormones is therefore determined by the presence of specific hormone *receptors* at the cell surface and by specific mechanisms within the cell that react to increased cAMP concentration.

As an example the actions of the hormones ACTH and epinephrine are, in general, very different. **ACTH**, from the adenohypophysis, stimulates steroid release in adrenal cortical cells whereas epinephrine, from the adrenal medulla, stimulates glycogen catabolism in hepatic cells. Actually, in their specific biochemical role, each hormone stimulates adenyl cyclase activity at the surface of its target cells, thus increasing the concentration of cAMP within the target cells. The actions of the two hormones differ because the liver cells have receptors that respond to epinephrine, whereas the adrenal cortex cells have receptors that respond to ACTH. They also differ because the mechanisms within liver cells respond to increased cAMP concentration by more rapid glucose breakdown, whereas the mechanisms within adrenal cortex cells respond to increased cAMP concentration by more rapid steroid production.

In the liver cell cAMP stimulates the phosphorolysis of glycogen to glucose 1-phosphate. What actually happens is that cAMP activates a protein *kinase*. When activated by phosphorylation the second kinase catalyzes the phosphorylation of the enzyme phosphorylase, which thus becomes activated as the catalyst for glycogen breakdown.

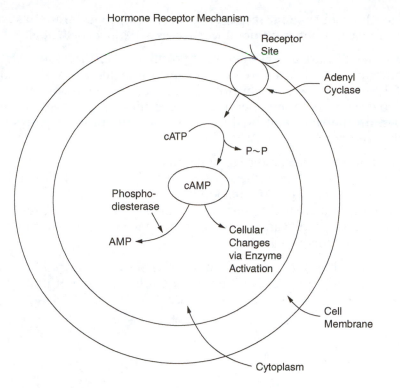

Hormone Receptor Mechanism

Recent research has shown that cAMP is active in almost every type of mammalian cell, affecting many different types of biochemical reactions. In many cases cAMP acts to modify enzyme activity, as we see in the liver cell, while in other cellular parameters cAMP stimulates synthesis of specific proteins. In some of the latter cases inhibitors of RNA synthesis block the stimulating effect of cAMP, indicating that cAMP affects mRNA transcription. In other cases the stimulating effect of cAMP is unaffected by inhibitors of RNA synthesis, indicating that the cAMP affects post-transcriptional processes.

The evidence at present suggests strongly that cAMP is not the only mechanism by which the total spectrum of hormones influence the cell. Steroids, thyroxin, GH, and insulin operate independently of the cAMP concentrations. You will recall that the synthesis of a hormone, whether it be protein or steroid, is controlled via mRNA to the ribosome. Thus, we can say that the production of hormones by cells is under the control of the DNA sequence known as a gene.

Current research has demonstrated that the insulin molecule (synthesized by the pancreas) attaches to specific membrane receptor sites but does not stimulate adenyl cyclase activity. Keep in mind that **insulin** is a relatively large polypeptide that was synthesized by the ribosomes of pancreatic cells. Insulin remains hormonally active even when it is bound to an inert polysaccharide carrier so large that the insulin molecule is prevented from entering the target cell. Thus, the actual mechanism of this and other hormones centers around the presence of cell membrane receptor sites and a very elaborate cytoplasmic enzyme transfer system. Other protein hormones are GH, TSH, FSH, LH, Prolactin, Oxytocin, ADH, Calcitonin, T_3, Parathyroid and Glucagon.

III. HORMONES (STEROIDS)

Because of their importance in animal medicine the steroid hormones have aroused a great deal of scientific interest. The principle steroid hormones of mammals

---

are the sex hormones (**androgens, estrogens,** and **progesterone**) as well as the adrenal cortical hormones (**mineralocorticoids** and **glucocorticoids**).

Steroids generally have a molecular weight of about 300. All the steroids are based on a common central structure of four interconnected rings of carbon atoms; three of the rings have six carbon atoms and the other ring has five atoms. The differences between the various steroids are determined by the pattern of chemical bonds within the rings and by the nature and orientation of the side groups attached to the rings.

The structural similarities of the steroid hormones are reflected in their origin. All of them are synthesized from the same chemical precursor, cholesterol. Although cholesterol has no known hormonal function, it is a sterol and has the characteristic four-ring structure. The hormones are synthesized by altering the side groups of cholesterol; the modifications are made by enzymes in the specific cells of tissues classed as endocrine glands.

Since the steroid hormones are small molecules, they can easily diffuse into many kinds of cells and just as easily diffuse back out if they are not utilized. They are effective even when in very small quantities.

Base Structure—Steroid

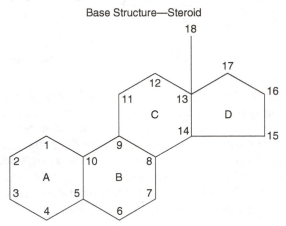

STEROID—applied to members of a group of compounds which have in common the cyclopentanoper-hydrophenonthrene nucleus.

Steroid hormones are always thought of in light to which effects they have or activities they influence with respect to specific target tissues. For example, the reproductive tract is an obvious target for the action of the androgens and estrogens. Recent research has shown that other tissue areas, liver, nervous system, and kidneys are also influenced by the specific type of steroids known as the sex hormones.

Adrenal glucocorticoids usually are regarded as regulators of carbohydrate and protein metabolism in various tissue compartments. However, it has been known for a long time that the glucocorticoids also form a very important part of the biological feedback loop by suppressing ACTH-RF release from the hypothalamus. Glucocorticoids also induce lysis of lymphoid cells, inhibit growth of fibroblasts and of regenerating liver cells, promote development of the pancreas and certain parts of the nervous system during embryonic life, stimulate GH synthesis by the adenohypophysis, and enhance adipose tissue catabolism with the subsequent release of free fatty acids into the extracellular fluids.

Aldosterone, the most studied of the mineralocorticoids, helps in the regulation of HOH and ion metabolism at the level of the kidney. It influences Na^+ transport across the nephron tubule in the kidney.

It appears that steroids ultimately influence the transcription of specific mRNA. When the steroid arrives at the cell membrane the first step is the formation by noncovalent bonds of a complex between the hormone and a specific protein receptor molecule (cytoplasm).

The simplicity in size of steroids does allow them access to the cytoplasm and thus the cell organelle membrane surfaces within the cell. You must remember that protein structured hormones are *generally* high molecular weight and *macromolecular* in nature. Because of these characteristics nonsteroid hormones influence cellular activities by activating cell membrane receptor sites.

IV. HORMONES (AMINES)

The cells of the adrenal medulla release two hormones, both examples of amines (catecholamines). They are **epinephrine** and **norepinephrine**.

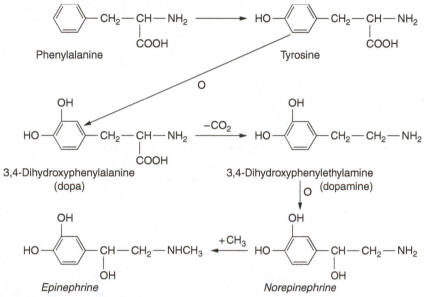

- These two amines are very similar in structure.
- The catecholamines disappear very rapidly from circulation and thus they must be released constantly from the adrenal medulla.

Give the anatomical location for the endocrine glands listed. Then list the hormones secreted, if they are protein or steroid hormones, and their target organ.

		Hormone	Type	Target Organ
A) Anterior pituitary				
	1.			
	2.			
	3.			
	4.			
	5.			
	6.			

		Hormone	*Type*	*Target Organ*
B) Posterior pituitary				
	1.			
	2.			
C) Thyroid				
	1.			
	2.			
D) Parathyroid				
	1.			
E) Adrenal Medulla				
	1.			
	2.			
F) Adrenal Cortex				
	1.			
	2.			
	3.			
G) Pancreas				
	1.			
	2.			
H) Thymus				
	1.			
I) Testes				
	1.			
J) Ovaries				
	1.			
	2.			
K) Pineal				
	1.			

Locate and label the endocrine glands: **thymus, thyroid, pineal, parathyroid, pituitary, pancreas, adrenals, ovaries, testes**

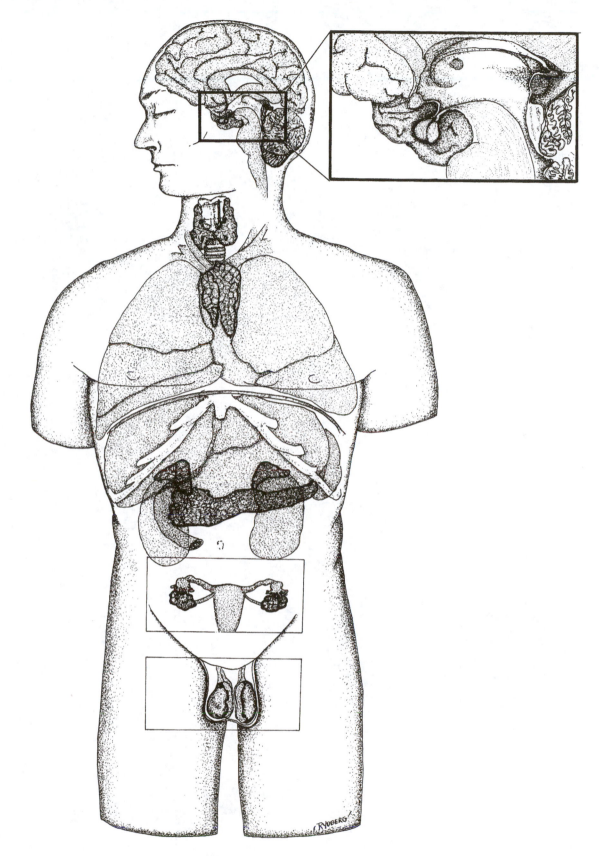

HUMAN, ENDOCRINE SYSTEM

REVIEW QUESTIONS

Using your lab, textbook, and any other materials available to you answer the following questions.

1. All hormones are steroid, protein, or amine. _____ (True or False)

MATCH

2. Insulin   ___   a. Protein

3. ADH   ___   b. Steroid

4. Testosterone ___   c. Amine

5. Glucagon   ___

6. FSH   ___

7. Estradiol   ___

8. TSH   ___

9. Epinephrine   ___

10. Thyroxine   ___

11. Only macromolecular protein hormones utilize cell membrane receptor

    sites. _____ (True or False)

12. Receptor sites are only associated with chemical molecules that have a

    molecular weight over 10,000. _____ (True or False)

13. Steroids generally have a molecular weight around 300.

    _____ (True or False)

14. The initial precursor of all steroids is acetate cholesterol.

    _____ (True or False)

15. TSH is a macromolecular protein. _____ (True or False)

16. Adrenalin (epinephrine) is a relatively small amine produced by the adrenal

    cortex. _____ (True or False)

17. The adrenal gland synthesizes both steroids and amines.

    _____ (True or False)

18. The function of the endocrine system is to integrate, correlate,
    and control body processes (metabolism and homeostasis).

    _____ (True or False)

19. List four ways the endocrine system may control or influence the rate and quantity of hormone secretion. Give one hormone example of each way.

    a. _____

    b. _____

    c. _____

    d. _____

20. Homeostasis will be maintained only when all tissues, including endocrine, make an acceptable contribution to the overall activity of the individual.

    _____ (True or False)

21. The ability to fully understand the role of hormones requires one to integrate

    the function of the various systems. _____ (True or False)

22. The hypothalamus is located in a skull depression known as the sella turcica.

    _____ (True or False)

23. The posterior pituitary (neurohypophysis) is located directly below the thalamus.

    _____ (True or False)

24. The hypothalamus is the site of ADH production. _____ (Yes or No)

25. The neurohypophysis is the site of ADH storage. _____ (Yes or No)

26. ADH aids, at the level of the kidney, in the retention of water. _____ (True or False)

27. ADH is a large macromolecular protein. _____ (True or False)

28. Iodine is utilized in the synthesis of the thyroid hormone (thyroxine).

    _____ (True or False)

29. There are receptor cells in the kidney and the thyroid for TSH.

    _____ (True or False)

30. Where are FSH, ACTH, and TSH synthesized? _____

31. How do ADH and oxytocin pass from their area of synthesis to their area of

    storage? _____

    _____

32. Steroids are produced by the ovary, testes, and adrenal cortex.

    _____ (True or False)

33. The pituitary (hypophysis) is attached to the hypothalamus by the

    infundibulum. _____ (True or False)

34. Current research has demonstrated that the pituitary is not the master gland.

    _____ (True or False)

35. The adrenal gland is located on the top of the kidney. _____
    (True or False)

36. The portion of the adrenal gland that completely surrounds it is called the

    _____.

37. The adrenal gland is divided into two regions, the _____ also
    known as

    the _____ _____, and the _____
    (middle).

38. The medulla releases the hormone _____ after nerve
    innervation.

39. The fight or flight syndrome starts with the cortex releasing adrenalin into

    general circulation. _____ (True or False)

40. The materials released by the cortex include the _____,

    _____, and _____.

41. Materials released by the medulla enter into circulation quickly, whereas

    materials released by the cortex may take hours. _____
    (True or False)

42. An example of a mineralocorticoid is _____.

43. An example of a glucocorticoid is _____.

44. An example of a sex hormone is _____.

45. Name the three types of hormones. _____,

    _____, and _____

46. Epinephrine is the chemical name for _____.

47. The thyroid is very vascular. _____ (True or False)

48. What kind of epithelial cells comprise the thyroid? _____

49. Calcitonin is an antagonist to what hormone? _____

50. Where is PTH produced? _____

51. We regularly deposit to and remove from bone Ca^{+2} and PO_4^{-3}.

     _____ (True or False)

52. TSH is produced by the _____.

53. You can live without the thyroid. _____ (True or False)

54. You can live without the parathyroid. _____ (True or False)

55. The pancreas is a dual-functioning structure, made up of both exocrine and

     endocrine cells. _____ (True or False)

56. The pancreas is housed in the first loop of the jejunum. _____
     (True or False)

57. The cells of the pancreas that produce insulin are termed the islets of

     _____.

58. What is the chemical nature of insulin? _____

59. How many amino acids comprise insulin? _____

60. Which pancreatic hormone converts glycogen to glucose?

     _____

61. On the average we store about _____ days worth of glycogen
     in our liver.

Reproductive System

I. INTRODUCTION

In this exercise, you will observe the details of the vertebrate reproductive system. This system is closely associated with the urinary system, and the two are often referred to as the **urogenital system.** We will concentrate our efforts on the reproductive portion at this time.

In a previous laboratory, you learned that the cell divisions leading to the production of **gametes**, **meiosis**, occur in specialized glands, collectively termed the **gonads**. At the end of this exercise you should develop an appreciation for the complexity of the systems which ensure that the male and female gametes unite to form a fertilized **zygote**. This is an important point! Everything about the structure and function of the reproductive tracts is fine-tuned for the process of fertilization, perpetuation of the human species. The details of the systems are fascinating, and you should develop an acute awareness of how these systems operate. There is more misunderstanding and downright ignorance about the functioning of our reproductive tracts, than for any other system. Hopefully, this exercise will help you understand how your reproductive system operates, as well as that of individuals of the opposite sex. The major objective of this exercise will be to develop an appreciation for the reproductive processes that occur in your own bodies and other individuals.

II. HUMAN FEMALE REPRODUCTIVE SYSTEM

The female gonads, the **ovaries**, are almond-sized, paired organs suspended from the wall of the peritoneum. These are the female gonads. Carefully examine the ovary and note a coiled tube beside it. This is the **oviduct** or **Fallopian tube.** This structure does not attach to the ovary, but begins as a small funnel which partially envelopes the ovary. **Ova** released from the ovary pass through the coelom, into this funnel, and down the oviduct. Most successful fertilizations take place in the oviduct.

Now you will locate the **uterus** which appears as a muscular-inverted triangle or a pear. The uterus "houses" the embryo and fetus during development. It is capable of distending from a volume of 3 to 4 teaspoons (nonpregnant) to 5–6 quarts (pregnant). Strong muscular contractions in the uterus are a signal of impending birth and these contractions assist the delivery of a newborn. Also, the uterus is the source of the monthly menstrual discharge in nonpregnant women.

Locate and label: ovaries, fallopian tube, uterus, cervix, vaginal canal, urethera, rectum

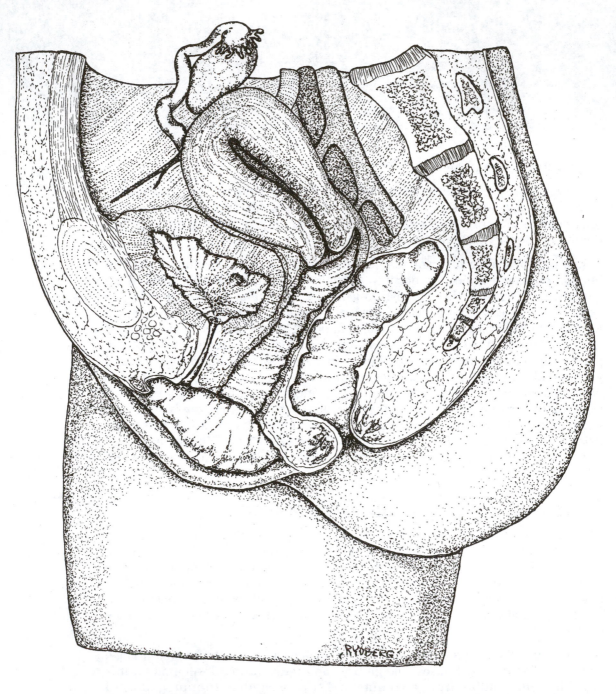

MEDIAN SAGITTAL SECTION

Figure 19.1   Reproductive system of the human female.

The **urethra** leaves the urinary bladder and passes downward. It carries urine. Just under the urethra, you should see the **vagina**. The vagina receives sperm during copulation and serves as the lower portion of the birth canal during parturition (birth). Under the vagina, you should note the **rectum** passing from the digestive tract.

The female external genitalia collectively are termed the **vulva** and consist of the **mons veneris** (a fatty pad lying over the pelvic bones, named for the Roman goddess of love), the **labia majora**, the **labia minora,** and the **clitoris**. These two pairs of labia are the outer and inner "lips" surrounding the vagina. The clitoris, homologous to the male's penis, is composed of spongy tissue and becomes erect and is a source of pleasure during sexual stimulation.

III. HUMAN MALE REPRODUCTIVE SYSTEM

The **testes**, male **gonads**, begin their development posterior to the kidneys in mammals. Just before birth, they descend into the paired **scrotal sacs**, which appear as outpocketings near the anus. The scrotal sacs are connected to the abdominal cavity by openings called the **inguinal canals.**

Alongside and attached to the testis is a highly-coiled tube, the **epididymis**. This structure stores sperm produced in the testis. Now, note a small tube leading from the epididymis and passing up through the inguinal canal. This is the **vas deferens**. Follow this structure. You will note that it leads from the epididymis, through the inguinal canal, loops over the umbilical arteries and the ureters (tubes leading from the kidneys to the urinary bladder), and connects just below the urinary bladder to the **urethra.**

The **urethra** is a tube which carries products of **both** the reproductive system in the male and urinary tract. In the male it leaves the urinary bladder then extends through the **penis**, the male intromittent organ. This structure is composed of erectile tissue and is capable of erection, sometimes in a matter of seconds. Cut through the skin just below the umbilical cord. This will expose the muscular penis. Note the following: **Cowper's gland** lying on either side of the urethra back near the anal opening; **prostate gland** located on the dorsal side of the urethra just below the junction of the urinary bladder and the urethra (this is often difficult to locate); **seminal vesicles,** paired glands located on either side of the prostate gland. The Cowper's glands, prostate, and seminal vesicles are called **accessory glands,** and provide fluid and sugars (**seminal fluid**) to the sperm as they exit through the urethra. Combined, the seminal fluid and sperm are called **semen**.

IV. MICROSCOPIC ANATOMY OF THE TESTES

Carefully study Figure 19.3 which presents increasing levels of detail of male gonads, the testes (singular testis or testicle). The testes have dual functions: production of male gametes, **spermatozoa**, and male sex hormones (**androgens**) such as **testosterone**. The latter produces and maintains the typical male phenotypic characteristics.

Now examine a prepared slide showing a cross section of a testis obtained from an adult mammal. Under low magnification, note that the viewing field consists mainly of many circular structures, the seminiferous tubules. Each human testis contains about 800 **seminiferous tubules.** The tubules are about the diameter of a coarse thread and their combined length is about 800 meters—nearly 1/2 of a

Locate and label: testes, scrotum, urethra, bladder, prostate, cowpers gland, epididymis, vas deferens, rectum, penis

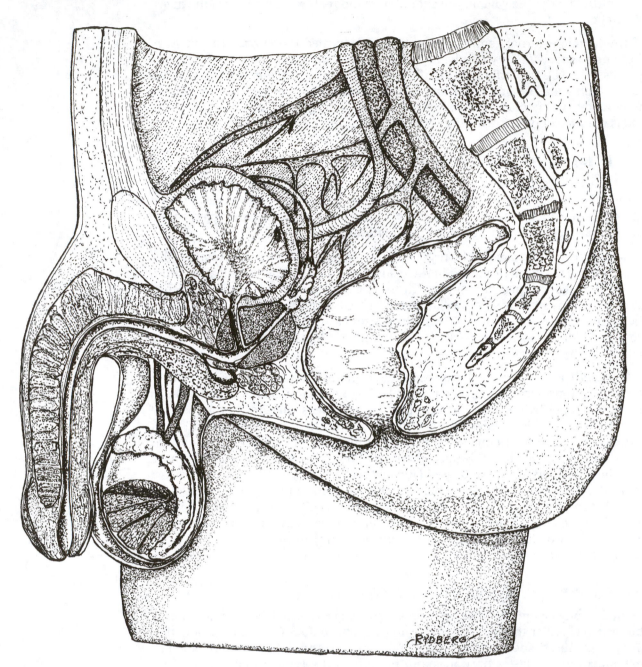

MEDIAN SAGITTAL SECTION

© Kendall/Hunt Publishing Company

Figure 19.2   Reproductive system of the human male.

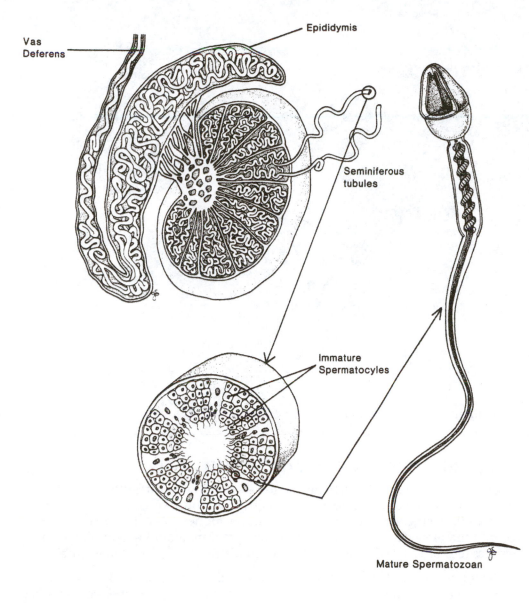

Figure 19.3   Increasing detail of the human testis.

mile! It is within the seminiferous tubules and through the process of meiosis, that spermatozoa are produced. Between the circular tubules are small masses of **interstitial cells** which produce testosterone.

Observe one tubule under high power. Locate mature sperm, which look like fine dark lines in the middle of the tubule. Sperm are among the smallest human cells. A dense packet of 100,000 sperm is barely visible to the "naked" eye. Males produce sperm from the time **puberty** is attained (about 12 to 14 years of age) until death. Healthy mature males produce some 300 to 500 million sperm daily. Models of a testis are available for your examination.

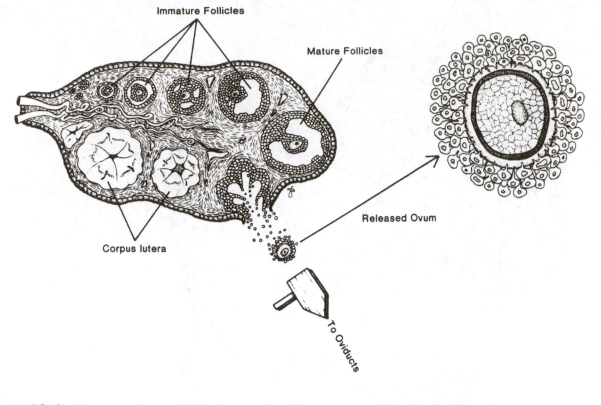

Immature Follicles

Mature Follicles

Released Ovum

Corpus lutera

To Oviducts

Figure 19.4 Ovary and ovum in the human female.

V. MICROSCOPIC ANATOMY OF THE OVARY

Carefully study Figure 19.4 showing a cross section of an ovary, the female gonad. Like the testes, ovaries perform two major functions. Ovaries release female gametes (ova, singular ovum) and female sex hormones. The latter play important roles in the menstrual cycle and in producing and maintaining feminine characteristics (e.g. breast development, fat and hair distributions, and skin texture).

Having inspected the above diagram, now examine a prepared cross-section view of an adult ovary. Under low magnification, note that the inner core consists of loose connective tissue while around the perimeter, circular-like structures are visible. These are the **ovarian follicles** which are the sites of ova production and maturation.

Locate a mature follicle and examine it under high power. The developing ovum is surrounded by cells within the liquid-filled follicle. At birth, a human female's ovaries may contain a million or more follicles; however, this number is reduced to about 300,000 by puberty. In the normal female, one mature ovum is released each month from the time puberty begins (usually 11 to 12 years of age) for the next 30 to 40 years, unless interrupted by pregnancy, until the time of **menopause** (usually 47 to 49 years of age). This means that, of the 300,000 or so follicles, only about 400 release mature ova (ovulate), and hence have the potential to produce a new individual. The development of the remainder of follicles is arrested and they degenerate.

Models of ovaries with developing follicles are available for your examination.

REVIEW QUESTIONS

1. Complete the following chart to describe the differences between human female and male reproductive systems.

	Female	Male
Gonad	_____	_____
Duct from this gland	_____	_____
Duct passes to which structure?	_____	_____
Final structure in the tract	_____	_____

2. Provide a gland, structure, or term for each of the following:

_____ Site of fertilization

_____ Birth canal

_____ A single gland which produces seminal fluid

_____ Tube connecting the epididymis with the urethra

_____ Common tube in the male which transports semen and urine

_____ Tubules of the testis where meiosis occurs

_____ Collective term for ova and sperm

Use your text & lab book and any other materials in lab to answer the following questions.

3. The testes are housed in a sac-like structure, the _____.

4. All primary cells of the seminiferous tubule are diploid. _____ (True or False)

5. All mammalian males, excluding pathology or trauma, can produce viable sperm until death. _____ (True or False)

6. Secondary cells of the seminiferous tubule are haploid. _____ (True or False)

7. A sex-change operation changes one's sex chromosomes. _____ (True or False)

8. Only the ovary, testes, and liver are capable of meiosis. _____ (True or False)

9. Menopause is characteristic of all mammalian females. _____
   (True or False)

10. Each human ovary contains thousands of follicle sites. _____
    (True or False)

11. Which hypophysial hormone influences a follicular site to begin maturation?
    _____

12. Immediately after ovulation the follicular site becomes secretory and is
    called the _____.

13. What hormones are produced by #12? _____ and _____

14. How long does #12 remain functional if pregnancy does occur? _____

15. When #12 degenerates the area that scars is known as the _____
    _____.

16. Identical twins result from two separate ova. _____ (True or False)

17. The egg is haploid while the sperm is diploid. _____ (True or False)

18. Specifically, meiosis in the male is termed _____.

19. The egg contains _____ (number of) autosomes.

20. Specifically, meiosis in the female is termed _____.

21. Collectively #24 and #29 are termed _____.

22. Following ovulation the egg is released directly into the Fallopian tube.
    _____ (True or False)

23. The vaginal canal is acidic (pH) in nature. _____ (True or False)

24. The sperm is considered foreign by the female's immune system. _____
    (True or False)

25. Fertilization occurs in the uterus. _____ (True or False)

26. As the sperm move toward the Fallopian tube many are destroyed, thus
    reducing the total sperm count. _____ (True or False)

27. The caudal surface of the uterus contain fimbriae. _____ (True or False)

28. The sperm has a life-expectancy of about _____ hours.

29. The Fallopian tube is also known as the _____.

30. The human egg is, microscopically, larger than the sperm. _____
    (True or False)

31. Implantation occurs in about 12–24 hours following fertilization. _____ (True or False)

32. The doubling of cells, beginning with the zygote, is termed _____.

33. Which hormone maintains the endometrium lining during pregnancy? _____

34. Germ layers form about _____ weeks after fertilization.

35. The human embryo has a tail and gill slits. _____ (True or False)

36. Oxygen is transported within the fetus by way of its own red blood cells. _____ (True or False)

37. The _____ _____ acts as an organ of respiration by providing for the exchange of O_2 and CO_2 between the maternal and fetal circulation.

38. The rupturing of the follicle and the discharging of the egg is termed _____.

39. What portion of the Fallopian tube brushes the swollen follicle? _____

40. The egg, following ovulation, is about the size of a grain of sand. _____ (True or False)

41. The human female egg has a nucleus. _____ (True or False)

42. The Fallopian tube is lined with cells that are ciliated. _____ (True or False)

43. The inside diameter of the Fallopian tube is only about twice that of a human hair. _____ (True or False)

44. How long does it take the egg to travel the distance from the ovary to the uterus? _____

45. The egg of the human must join with the sperm in approximately _____ hours if fertilization is to occur.

46. What happens to the egg if there is no fertilization? _____

What sex chromosomes do every female and male cell contain? _____

What is the name applied to the area where the follicle ruptured and becomes secretory? _____ _____

What hormones are produced by this area? _____ and _____

From *General Biology 1407: Lecture and Laboratory Manual* by L. Jack Pierce. Copyright © 1998 by Kendall/Hunt Publishing Company. Used with permission.

47. Cowper's gland produces fluid that will aid the sperm on their journey to the exterior. _____ (True or False)

48. The fluid produced by several areas of the male reproductive tract is termed _____.

49. Sperm, in man, are produced at the rate of _____ every 24 hours.

50. What is the optimum temperature of the testes for maximum sperm production? _____

51. When sperm are produced they are transported to the _____ for storage.

52. What is the chromosome number of the human sperm? _____

53. Approximately how long are mature sperm stored in the epididymis? _____

54. What is the name applied to the tail of the sperm? _____

NUTRITION AND METABOLISM COMMENTS

1. FALSE. Skinless roasted chicken thighs and legs have more fat than many types of steak (9% versus 5%). However, skinless roasted chicken or turkey breast are the best bet (3.5%). Ground or untrimmed red meats have a high fat content.

2. TRUE. Malnutrition may be due to a deficiency or a surplus of any or all of these foods and nutrients. Balance is the key to complete nutrition.

3. FALSE. When food is frozen rapidly this results in a minimal loss of vitamin C, with all other nutrients being maintained. Fresh foods, which have been exposed to the air in display cases, have fewer vitamins and minerals than many fresh frozen foods.

4. FALSE. They are just about equal. However, the fat in margarine is usually less saturated, which is recommended for individuals who should lower their intake of cholesterol. Nonetheless, the process of hydrogenation, which allows the vegetable oil in margarine to stay solid at room temperature, has been implicated in increased risk for certain types of cancer.

5. TRUE. A surplus of animal fats in the diet may result in an accumulation of cholesterol and fatty acids (triglycerides) which have been shown to have a direct link to arterial disease. Fat intake greater than 20% of total calories has been linked to increased incidence of many types of cancer including breast, skin, colon, and prostate cancer.

6. TRUE. The more times a person exercises each week the greater the reduction in risk. This is because exercise results in the reduction of fat tissue and makes cells more responsive to insulin.

7. FALSE. Moderation is the key in all dietetic behavior. Most vegetable oils are highly unsaturated fats. However, tropical oils, such as coconut and palm oil, have been shown to be very high in saturated fats.

8. FALSE. Cooking eggs protects the eater from possible *Salmonella* contamination.

9. FALSE. The brain is an organ of the body, and subsequently receives its nutrients in much the same way as do the rest of the body's organs. No individual foods specifically send their nutrients to any one body organ or area.

10. TRUE. As much as 10% of the population is allergic to one or more proteins in certain foods. Symptoms may include eczema, hives, itchy skin, asthma, vomiting, diarrhea, and other symptoms.

11. FALSE. Individuals with a fever or a cold require good nutrition.

12. FALSE. Many athletes, including some world champions and professionals, either are or have been vegetarians. Proper selection is the key. Select any

From *Human Biology Laboratory Manual,* Fourth Edition/Revised Printing by Keith Cunningham and Leslie Snider. Copyright © 2001 by Kendall/Hunt Publishing Company. Used with permission.

two vegetable proteins (such as rice and beans) and you will normally obtain all the essential amino acids.

13. FALSE. Analysis has shown the same levels of nutrients in plants which have been grown by other means, including hydroponics.

14. FALSE. Water contains no calories and therefore cannot be fattening. However, water may provide the illusion of gained weight due to water retention in the tissues of some individuals.

15. FALSE. The proper functioning of the reproductive system depends upon a balanced diet.

16. TRUE. Exercise in conjunction with a balanced low calorie diet can lead to long term weight loss. If you diet without exercising you may lose muscle mass as well as fat. Additionally, exercise appears to raise your basal metabolic rate which helps prevent subsequent weight gain.

17. FALSE. People generally skip breakfast, which can be the most important meal of the day. This usually leads to increased snacking.

18. FALSE. Vitamins primarily act as co-enzymes and catalyze reactions, but they cannot take the place of carbohydrates, fats, and proteins. They simply enable the body to utilize these foods.

19. FALSE. Raw sugar contains negligible amounts of vitamins and minerals.

20. FALSE. Due to a decrease in basal metabolic rate as we grow older, there should be a gradual reduction in the caloric intake of an individual as he or she ages.

21. FALSE. Juices contain natural and added sugar and therefore have a large number of calories.

22. TRUE. Reducing under these circumstances can often be a lifesaving measure.

23. TRUE. For this reason many restaurants are careful to constantly change oils used in deep frying.

24. TRUE. Blood vessel damage may be caused by repeated losses followed by gains.

25. TRUE. Use as little water as possible in the preparation of fruits and vegetables. Any fluids left over should be used as juice or in sauces. Microwaving greatly reduces loss of vitamins while cooking.

26. TRUE. There may be hidden problems. Most overweight people should have a physical examination before any weight-loss program is started.

27. TRUE. Additional weight puts added strain upon the heart and lungs.

28. TRUE. A breakfast which contains a starch source such as potatoes, rice, cereals, or breads results in a gradual release of energy that lasts until lunch. This results in a longer lasting energy and reduces the need for stimulants which can cause fatigue later in the day.

29. FALSE. Although this is probably true for many individuals, there are some people who benefit most from 4 to 6 small meals a day.

30. FALSE. These foods are no more constipating than other foods. A primary cause of constipation is stress.

31. TRUE. These can therefore be used as a good source of nutritional information. Be observant, however, as there are many means of disguising total sugar content by listing sugar under a variety of different names.

32. TRUE. These foods, as well as others such as alcohol, caffeine, chocolate, and fermented foods, cause blood vessels in the head to expand. The oversupply of blood increases pressure and brings on headaches.

33. FALSE. More weight is lost if a diet limits both fat intake *and* total calories consumed.

34. TRUE. However, you need to fine tune the amount to account for body size, fitness and activity level, diet, and climate. The best way to determine if you

From *Human Biology Laboratory Manual,* Fourth Edition/Revised Printing by Keith Cunningham and Leslie Snider. Copyright © 2001 by Kendall/Hunt Publishing Company. Used with permission.

are drinking enough water is to monitor the color of your urine. If it is a pale-yellow you are getting enough fluid for proper waste excretion.

35. TRUE. Death is the most graphic way that nature has of telling you to slow down your life style. Heed the warning signs of death: lack of breathing; extreme loss of appetite; lack of all sensations; poor, if any, circulation; general BLAH feeling; absolutely no sex drive.

NUTRIENT ALLOWANCES AND CALORIE VALUES

One simple method used to calculate daily nutrient allowances (grams) is demonstrated below.

Daily allowance of nutrients (grams)	=	recommended % of total daily calories obtained from nutrient	×	total number of calories ingested each day	−	number of calories supplied per gram of nutrient

Examples:						
Total Fats (grams)	=	0.30 (30%)	×	2000 calories/9 calories/g	=	66.7 g
		0.30 (30%)	×	2500 calories/9 calories/g	=	83.3 g
Cabohydrates (grams)	=	0.60 (60%)	×	2000 calories/4 calories/g	=	300 g
		0.60 (60%)	×	2500 calories/4 calories/g	=	375 g
Protein (grams)	=	0.8 g (10%) of protein/kg of body weight for 58 kg person				
		0.8 g/kg × 58 kg = 46 g of protein/day				

CALORIE VALUES

Food	Calorie
Apple, 1 fresh	115
Apple juice, 1 cup	120
Apple pie, 1/6 of 9" pie	380
Applesauce, 1/2 cup	50
Asparagus, 8 stalks	25
Bacon 2 crisp 6" strips	100
Banana, 1 medium sized	130
Beans, baked, 1/2 cup	160
Beans, green, 1/2 cup cooked	15
Beans kidney, 1/2. cup cooked	115
Beans, lima, 1/2 cup cooked	80
Beef, corned, 3 oz.	220
Beef, filet mignon, 4 oz.	400
Beef, hamburger, large patty	300
Beef, roast (slice)	96
Beef, sirloin, steak 3 oz.	300
Beef stew 1 cup	260
Beer, 12 oz.	170
Beets, 1/2 cup	40
Berry pie, 1/6 of a pie	370
Blackberries, 1/2 cup	40
Blueberries, 1/2 cup	45

Food	Calorie
Bologna, 1 slice	65
Bread, rye, 1 slice	55
Bread, white, 1 slice	60
Bread, whole wheat, 1 slice	55
Broccoli, 2/3 cup	30
Butter, 1 pat	50
Cabbage, 1/2 cup, raw	10
Cantaloupe, 1/2 of a melon	30
Carbonated water	0
Carrots, 1/2 cup, cooked	30
Catsup, 1 tablespoon	20
Celery, raw 3 small inner stalks	10
Cheese, American, 1" cube	80
Cheese, cottage, 1/2 cup	100
Cheese, Parmesan, 1 tbsp., grated	20
Cheese, Roquefort, 1" cube	40
Cheese, Swiss, 1 oz.	105
Cherries, fresh, sweet, 15 large	60
Chicken, broiler 1/2 medium	125
Chicken, fried, 1/2 breast	230
Chicken, roasted, 1 slice, dark	170
Chicken, roasted, 1 slice, light	165
Chicken salad, 1/2 cup	200
Chocolate, 1 cup made with milk	280
Coffee, clear, 1 cup	0
Coffee, w/1 tbsp. cream 1 cup	30
Coffee w/1 lump sugar 1 cup	30
Cola beverages, 1 glass, 8 oz.	105
Cookies, sugar, 1 3" diameter	65
Corn, canned, 1/2 cup drained solids	70
Corn flakes, 3/4 cup	60
Corn on cob, 1 medium	90
Cracker, graham, 1 2-1/2" square	15
Cream, whipping, 2 tbsp.	100
Cream of wheat, cooked, 3/4 cup	100
Cucumber, 12 slices	10
Doughnut, sugared 1	150
Egg, 1 fried w/1 teaspoon butter	105
Egg, 1 whole, boiled or poached	75
Eggplant, 2 slices, breaded & fried	210
Farina, 3/4 cup cooked	100

Food	Calorie
Grapefruit, 1/2 medium	70
Grapes, green seedless, 60	65
Green pepper, 1 whole	20
Ham, baked medium, 1 slice	200
Honey, 1 tablespoon	60
Honeydew melon, 1/4 medium-sized	30
Ice cream, plain, 1/2 cup	150
Jam or jelly, 1 tablespoon	60
Lemonade, 1 cup	100
Lentil soup, 1 cup, home-made	605
Lettuce, iceberg, 1/4 large head	20
Lettuce, 6 large leaves	20
Lobster, fresh, 1 cup	105
Macaroni, cooked, plain 1/2 cup	110
Macaroni & cheese, 1/2 cup	300
Maple syrup, 1 tablespoon	50
Margarine, 1 tablespoon	100
Mayonnaise, 1 tablespoon	110
Meat loaf, beef & pork, 1 slice	265
Milk, half and half, 1/2 cup	165
Milk, skimmed, 1 cup	90
Milk, whole fresh, 1 cup	165
Muffin, English, 1 large	280
Mushrooms, fresh 10 small	15
Oatmeal, cooked, 1/2 cup	75
Olive oil, 1 tablespoon	125
Olives, green 2 small or 1 large	20
Olives, ripe, 2 small or 1 large	25
Orange, 1 average-sized	75
Orange ice, 1/2 cup	110
Orange juice, 1 cup	110
Pancake, 1 4" diameter	75
Parsley, chopped, 2 tablespoons	2
Peaches, 2 halves w/2 tbsp. juice	85
Peanut butter, 2 tablespoons	180
Peanuts, 20 to 24 nuts	100
Pears, 2 halves w/2 tbsp. juice	85
Pear, fresh, 1 medium-sized	60
Peas, canned, 1/2 cup	75
Pickles, cucumber, 1 sweet-sour	20

Food	Calorie
Pineapple, canned, 1 slice w/juice	80
Popcorn, 1-1/2 cups, no butter	100
Pork chop, rib, broiled, 1	290
Pork roast, 4 oz.	450
Pork tenderloin, 2 oz.	200
Potato, baked, 1 medium-sized	90
Potato, boiled, 1 medium-sized	90
Potato chips, 8 to 10 large	100
Potato, sweet baked, 1 medium-sized	155
Pretzels, 5 small sticks	20
Pumpkin pie, 1/6 of 9" pie	300
Radishes, 5 medium	10
Raisins, seedless, 1/4 cup	110
Rice, brown, cooked, 1/2 cup	100
Rice, white, cooked, 1/2 cup	100
Roll, hard, white, 1 average-sized	95
Salmon, canned, 1/2 cup	205
Sausage, pork link, 3" × 1/2"	95
Shrimp, boiled, 5 large	70
Shrimp, fried, 4 large	260
Spaghetti, plain, cooked, 1 cup	220
Spinach, cooked & chopped, 1/2 cup	240
Split pea soup, 1 cup	270
Strawberries, fresh, 1/2 cup	30
Sugar, brown, 1 tbsp.	50
Sugar, granulated, 1 tbsp.	50
Tangerine, 1	35
Tea, clear, unsweetened, 1 cup	0
Tomato, fresh, 1 medium-sized	25
Tomato soup, clear, 1 cup	100
Tuna fish, water packed, 1/2 cup	165
Tuna fish, canned in oil, 1/2 cup	300
Turkey, roast, dark meat	205
Turkey roast, light meat	185
Vegetable soup, 1 cup	100
Waffles, 1, 5-1/2 dia.	230
Watermelon, 1 slice	90
Yogurt, whole milk, 1/2 cup	80
Zuchini, cooked, 1 cup	40

ANSWERS TO ENDOCRINE SYSTEM REVIEW QUESTIONS

Using your lab, textbook, and any other materials available to you answer the following questions.

1. All hormones are steroid, protein, or amine. ___True___ (True or False)

MATCH

2. Insulin __a__   a. Protein

3. ADH __a__   b. Steroid

4. Testosterone __b__   c. Amine

5. Glucagon __a__

6. FSH __a__

7. Estradiol __b__

8. TSH __a__

9. Epinephrine __c__

10. Thyroxine __a__

11. Macromolecular protein hormones utilize cell membrane receptor

    sites. ___True___ (True or False)

12. Receptor sites are only associated with chemical molecules that have a

    molecular weight over 10,000. ___False___ (True or False)

13. Steroids generally have a molecular weight around 300.

    ___True___ (True or False)

14. The initial precursor of all steroids is acetate cholesterol.

    ___True___ (True or False)

15. TSH is a macromolecular protein. ___True___ (True or False)

16. Adrenalin (epinephrine) is a relatively small amine produced by the adrenal

    cortex. ___False___ (True or False)

17. The adrenal gland synthesizes both steroids and amines.

    ___True___ (True or False)

18. The function of the endocrine system is to integrate, correlate,
    and control body processes (metabolism and homeostasis).

    ___True___ (True or False)

19. The posterior pituitary (neurohypophysis) is located directly below the thalamus.

    _____True_____ (True or False)

20. ADH aids, at the level of the kidney, in the retention of water.
    _____True_____ (True or False)

21. ADH is a large macromolecular protein. _____True_____ (True or False)

22. Iodine is utilized in the synthesis of the thyroid hormone (thyroxine).

    _____True_____ (True or False)

23. There are receptor cells in the kidney and the thyroid for TSH.

    _____False_____ (True or False)

24. Where are FSH, ACTH, and TSH synthesized? __anterior pituitary__

25. Steroids are produced by the ovary, testes, and adrenal cortex.

    _____True_____ (True or False)

26. The pituitary (hypophysis) is attached to the hypothalamus by the

    infundibulum. _____True_____ (True or False)

27. Current research has demonstrated that the pituitary is not the master gland.

    _____True_____ (True or False)

28. The adrenal gland is located on the top of the kidney. _____True_____
    (True or False)

29. The portion of the adrenal gland that completely surrounds it is called the

    ___adrenal cortex___.

30. The adrenal gland is divided into two regions, the _____cortex_____
    known as and the _____medulla_____ (middle).

31. The medulla releases the hormone ___epinephrine___ after nerve
    innervation.

32. The fight or flight syndrome starts with the cortex releasing adrenalin into

    general circulation. _____True_____ (True or False)

33. The materials released by the cortex include the __Glucocorticoids__,

    ___Androgens___, and __mineralcorticoids__.

34. An example of a mineralocorticoid is ___Aldosterone___.

35. An example of a glucocorticoid is _____cortisol_____.

36. An example of a sex hormone is _Estrogen (testosterone or progesterone)_.

37. Name the three types of hormones. _____protein_____,

_____amine_____, and _____steriod_____

38. Epinephrine is the chemical name for _____adrenalin_____.

39. The thyroid is very vascular. _____True_____ (True or False)

40. What kind of epithelial cells comprise the thyroid? ___simple cuboidal___

41. What is the chemical nature of thyroxine? _____protein_____

42. Calcitonin is an antagonist to what hormone? __parathyroid hormone__

43. Where is PTH produced?_parathyroid gland_

44. We regularly deposit to and remove from bone Ca^{+2} and PO_4^{-3}.

_____True_____ (True or False)

45. TSH is produced by the __anterior pituitary__.

46. You can live without the thyroid. _____True_____ (True or False)

47. You can live without the parathyroid. _____False_____ (True or False)

48. Today, we can destroy thyroid cells by ingesting radioactive iodine.

_____True_____(True or False)

49. The pancreas is a dual-functioning structure, made up of both exocrine and

endocrine cells. _____True_____ (True or False)

50. The pancreas is housed in the first loop of the jejunum. _____True_____
(True or False)

51. The cells of the pancreas that produce insulin are termed the islets of

_____langerhans_____.

52. What is the chemical nature of insulin? _____protein_____

53. How many amino acids comprise insulin? _____51_____

54. Which pancreatic hormone converts glycogen to glucose? _____Glucagon_____

55. On the average we store about _____1_____ days worth of glycogen
in our liver.

ANSWERS TO REPRODUCTIVE SYSTEM REVIEW QUESTIONS

1. Complete the following chart to describe the differences between human female and male reproductive systems.

	Female	Male
Gonad	ovaries	testes
Duct from this gland	fallopian tubes	epididymis
Duct passes to which structure?	uterus	vas deferens
Final structure in the tract	vagina	urethra

2. Provide a gland, structure, or term for each of the following:

   __fallopian tube__ Site of fertilization

   __vagina__ Birth canal

   __seminal vesicles__ A single gland which produces seminal fluid

   __vas deferens__ Tube connecting the epididymis with the urethra

   __urethra__ Common tube in the male which transports semen and urine

   __seminiferous tubules__ Tubules of the testis where meiosis occurs

   __gametes__ Collective term for ova and sperm

Use your text & lab book and any other materials in lab to answer the following questions.

3. The testes are housed in a sac-like structure, the _____scrotal sacs_____.

4. All primary cells of the seminiferous tubule are diploid. _____True_____ (True or False)

5. All mammalian males, excluding pathology or trauma, can produce viable

   sperm until death. __True__ (True or False)

6. Secondary cells of the seminiferous tubule are haploid. _____True_____ (True or False)

7. Immediately following spermatogenesis the sperm produced have only

   about a 2% capability of fertilization. __True__ (True or False)

8. A sex-change operation changes one's sex chromosomes. _____False_____ (True or False)

9. Only the ovary, testes, and liver are capable of meiosis. _____False_____ (True or False)

10. Menopause is characteristic of all mammalian females. _____True_____
(True or False)

11. Human females alternate the ovary-ovulation process. _____True_____
(True or False)

12. Each human ovary contains thousands of follicle sites. _____True_____
(True or False)

13. Which hypophysial hormone influences a follicular site to begin maturation?

_____FSH_____

14. Immediately after ovulation the follicular site becomes secretory and is

called the __corpus luteum__.

15. What hormones are produced by #17? ___Estrogen___ and __Progesteron__

16. Identical twins result from two separate ova. ___False___ (True or False)

17. The egg is haploid while the sperm is diploid. ___False___ (True or False)

18. Specifically, meiosis in the male is termed _____Spermatogenesis_____.

19. Nerve tracts, blood vessels, and the vas deferens enter and exit the body

cavity proper via the _____Inguinal_____ canal.

20. The egg contains _____22_____ (number of) autosomes.

21. Specifically, meiosis in the female is termed _____oogenesis_____.

22. Collectively #24 and #29 are termed ___gametogenesis___.

23. Following ovulation the egg is released directly into the Fallopian tube.

__False__ (True or False)

24. The vaginal canal is acidic (pH) in nature. ___True___ (True or False)

25. The sperm is considered foreign by the female's immune system. ___True___
(True or False)

26. Fertilization occurs in the uterus. ___False___ (True or False)

27. As the sperm move toward the Fallopian tube many are destroyed, thus

reducing the total sperm count. ___True___ (True or False)

28. The caudal surface of the uterus contain fimbriae. ___False___ (True or False)

29. The sperm has a life-expectancy of about _____48_____ hours.

30. The Fallopian tube is also known as the _____uterine tubes____.

31. The human egg is, microscopically, larger than the sperm. _____True_____ (True or False)

32. Implantation occurs in about 12–24 hours following fertilization. ___False___ (True or False)

33. The doubling of cells, beginning with the zygote, is termed ____cleavage____.

34. Growth and differentiation are one and the same. ___False___ (True or False)

35. Germ layers form about ___1–2___ weeks after fertilization.

36. The human embryo has a tail and gill slits. ___True___ (True or False)

37. Oxygen is transported within the fetus by way of its own red blood cells.

    ___False___ (True or False)

38. The ____placenta____ acts as an organ of respiration by providing for the exchange of O_2 and CO_2 between the maternal and fetal circulation.

39. With respect to relative size of body parts the fetus at 2 months has more area of the head and neck than arms and legs. ___True___ (True or False)

40. At birth the head and neck area has been reduced while the area of arms and legs has increased. ___True___ (True or False)

41. The opening between the upper chambers of the heart closes within 24 hours after birth. _False_ (True or False)

42. What portion of the Fallopian tube brushes the swollen follicle? _____fimbriae_____

43. The egg, following ovulation, is about the size of a grain of sand. ___False___ (True or False)

44. The human female egg has a nucleus. ___True___ (True or False)

45. The Fallopian tube is lined with cells that are ciliated. ___True___ (True or False)

46. The inside diameter of the Fallopian tube is only about twice that of a human hair. ___True___ (True or False)

47. How long does it take the egg to travel the distance from the ovary to the uterus? ____about 5–7 days____

48. The egg of the human must join with the sperm in approximately _____24_____ hours if fertilization is to occur.

49. What happens to the egg if there is no fertilization? ____it disintegrates____

50. Cowper's gland produces fluid that will aid the sperm on their journey to the exterior. ____True____ (True or False)

51. The ____Cowper's____ gland produces an alkaline fluid to protect the sperm.

52. The fluid produced by several areas of the male reproductive tract is termed _____semen_____.

53. The _____seminal_____ _____vesicles_____ releases a sugar solution to nourish the sperm.

54. What is the approximate length of the seminiferous tubules in man? _____1/2 mile_____.

55. Sperm, in man, are produced at the rate of ____300–500 million____ every 24 hours.

56. What is the optimum temperature of the testes for maximum sperm production? _____34°C_____

57. When sperm are produced they are transported to the ____epididymis____ for storage.

58. What is the chromosome number of the human sperm? _____23_____

59. Approximately how long are mature sperm stored in the epididymis? _____2–4 weeks_____

60. What is the name applied to the tail of the sperm? _____flagellum_____

FIGURES FOR SELF TEST

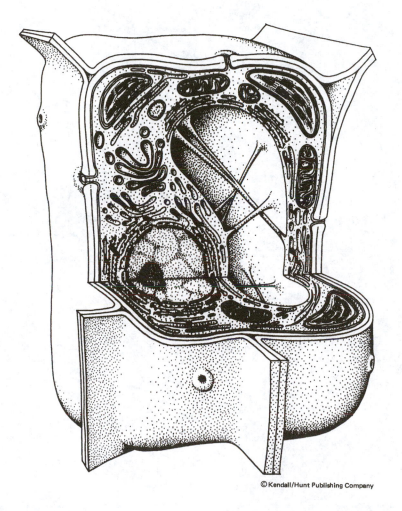

Generalized plant cell

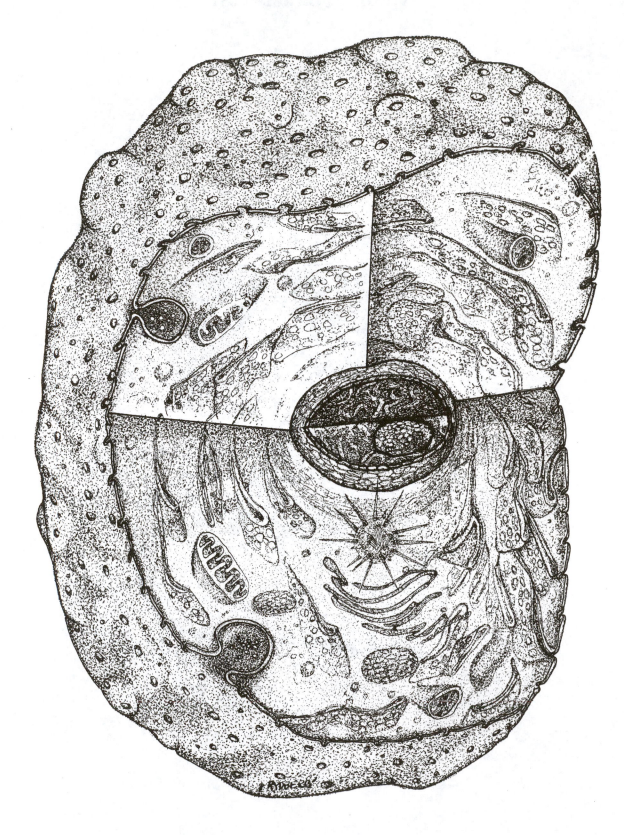

Animal cell diagram

From *Human Biology Laboratory Manual,* Fourth Edition/Revised Printing by Keith Cunningham and Leslie Snider. Copyright © 2001 by Kendall/Hunt Publishing Company. Used with permission.

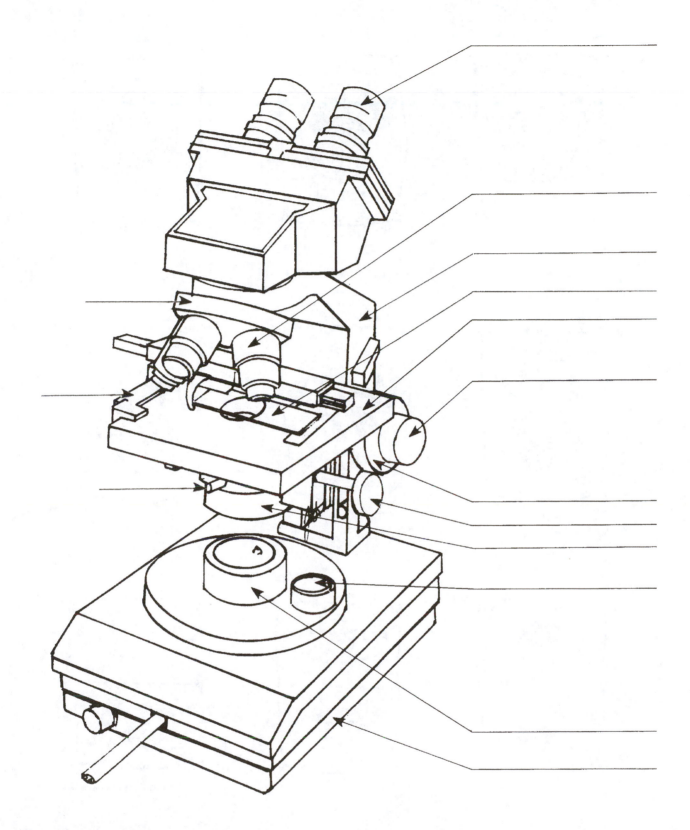

Schematic of a binocular compound microscope

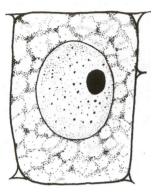
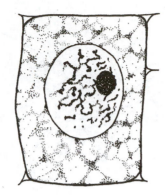

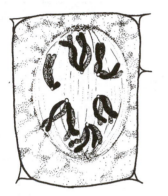

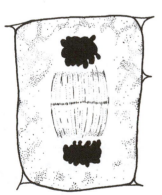
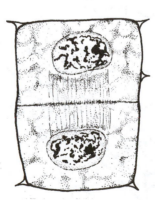
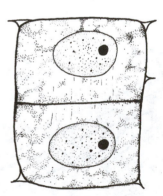

Mitosis in plant cells

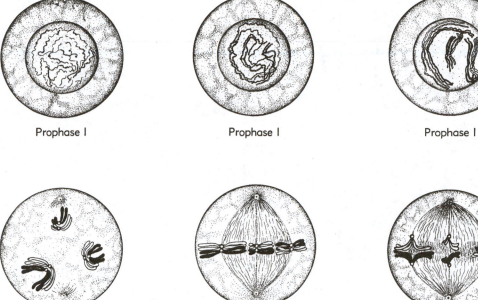

Prophase I

Prophase I

Prophase I

Prophase I

Metaphase I

Anaphase I

Telophase I
Prophase II

Metaphase II

Anaphase II

Telophase II

Meiosis in animals

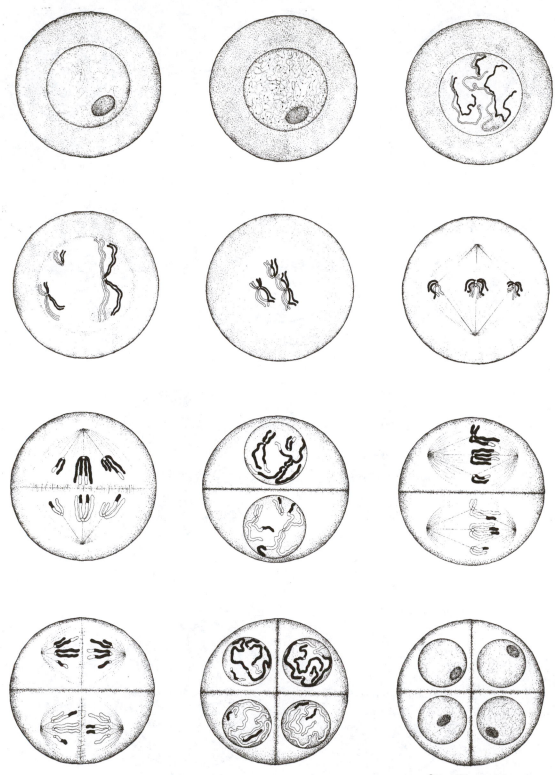

Meiosis in plants

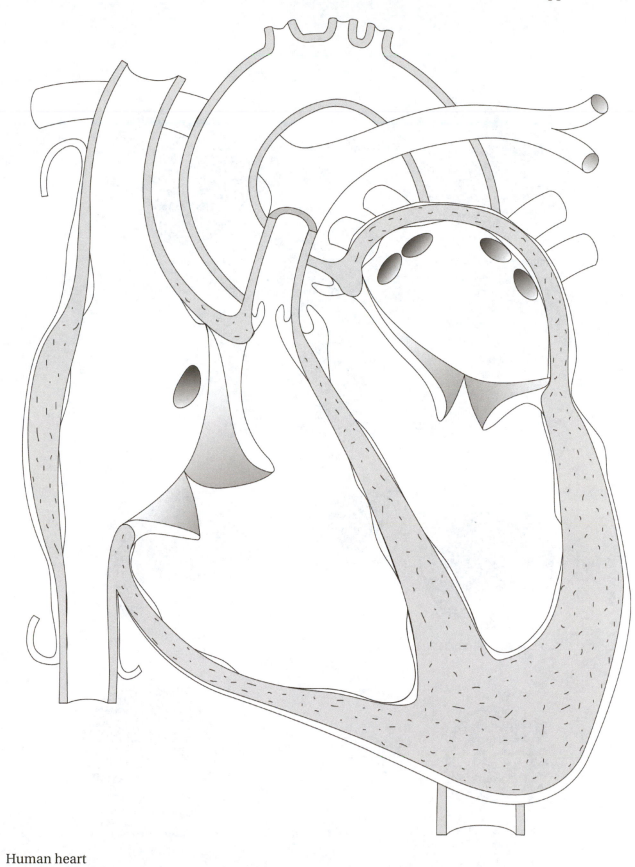

Human heart

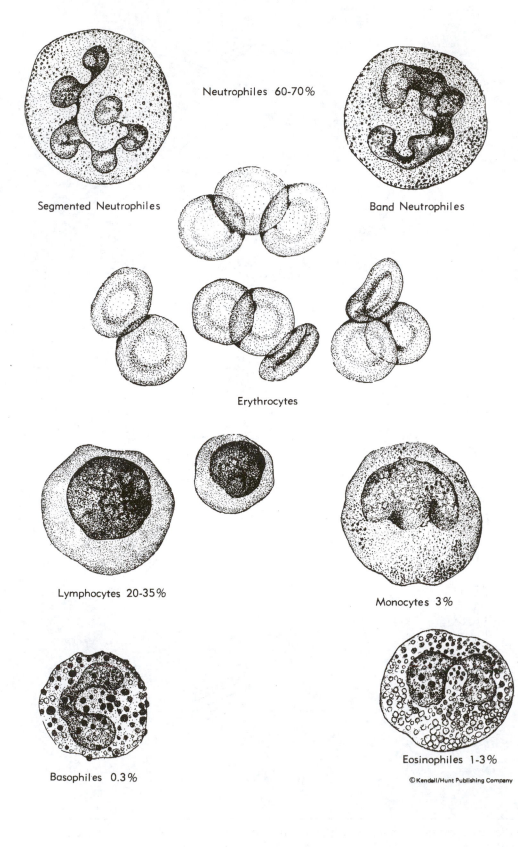

Neutrophiles 60-70%

Segmented Neutrophiles

Band Neutrophiles

Erythrocytes

Lymphocytes 20-35%

Monocytes 3%

Basophiles 0.3%

Eosinophiles 1-3%

Types of blood cells

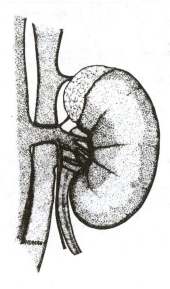

Figure A. External View

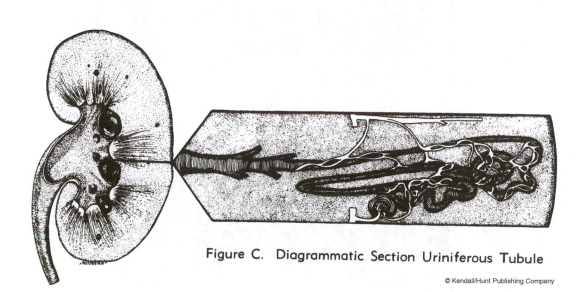

Figure C. Diagrammatic Section Uriniferous Tubule

Figure B. Sagittal View

Human kidney

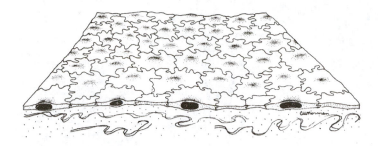

Figure A. Simple Squamous

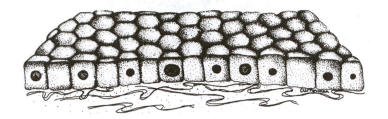

Figure B. Simple Cuboidal

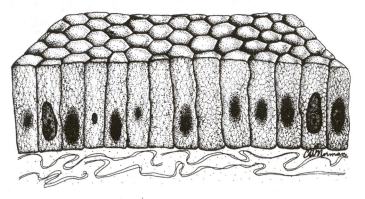

Figure C. Simple Columnar

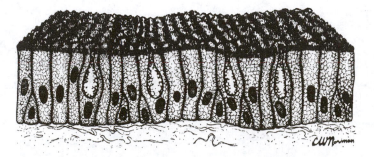

Figure D. Pseudostratified Columnar

Epithelial tissue cells

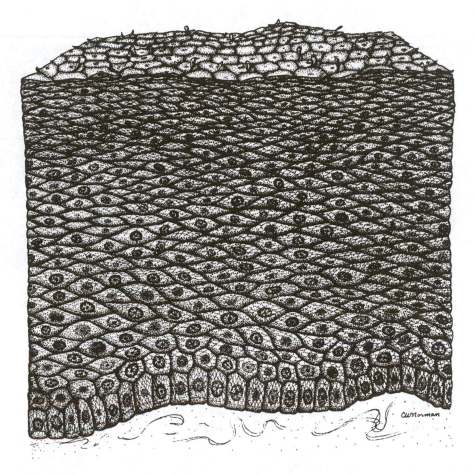

Figure A. Stratified Squamous Epithelium

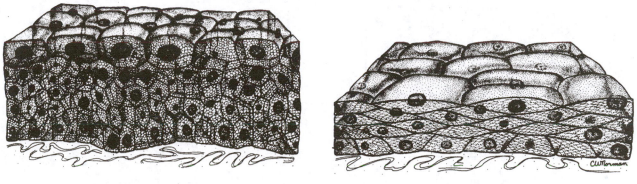

Relaxed

Extended

Figure B. Transitional Epithelium

Epithelial tissue cells

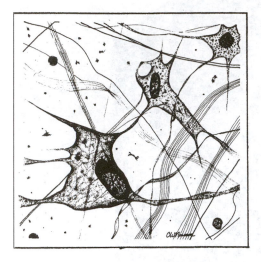

Figure A. Areolar

Figure B. Dense Fibrous

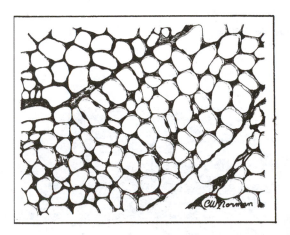

Figure C. Adipose

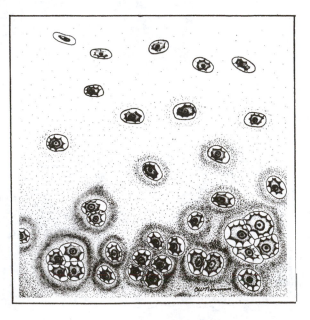

Figure D. Cartilage

Connective tissue cells

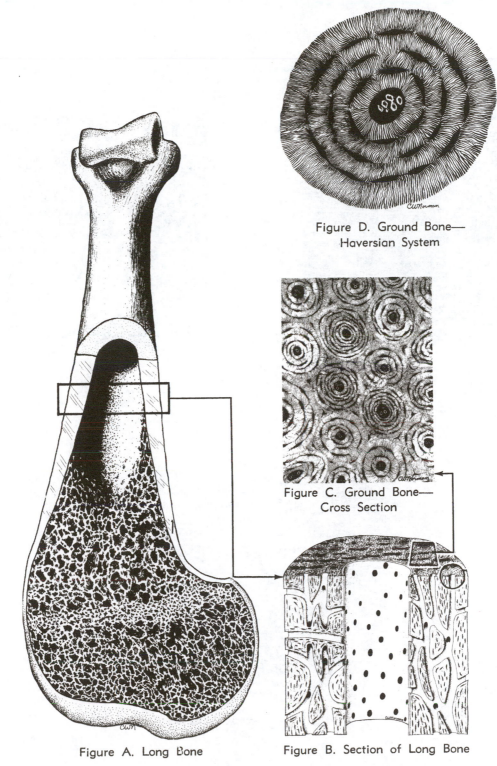

Figure D. Ground Bone—
Haversian System

Figure C. Ground Bone—
Cross Section

Figure A. Long Bone

Figure B. Section of Long Bone

© Kendall/Hunt Publishing Company

Bone tissues

Figure A. Smooth Muscle

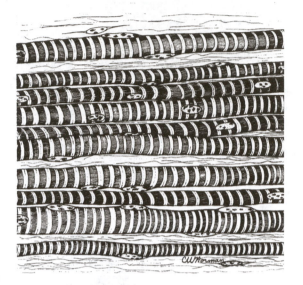

Figure B. Striated Muscle

Figure C. Cardiac Muscle

Muscle tissue cells

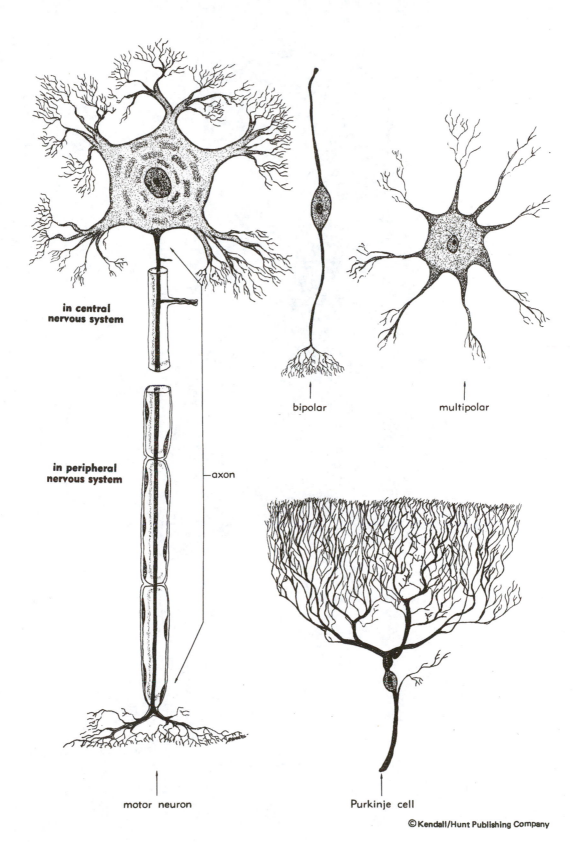

in central
nervous system

bipolar

multipolar

in peripheral
nervous system

axon

motor neuron

Purkinje cell

Nerve tissue, types of neurons

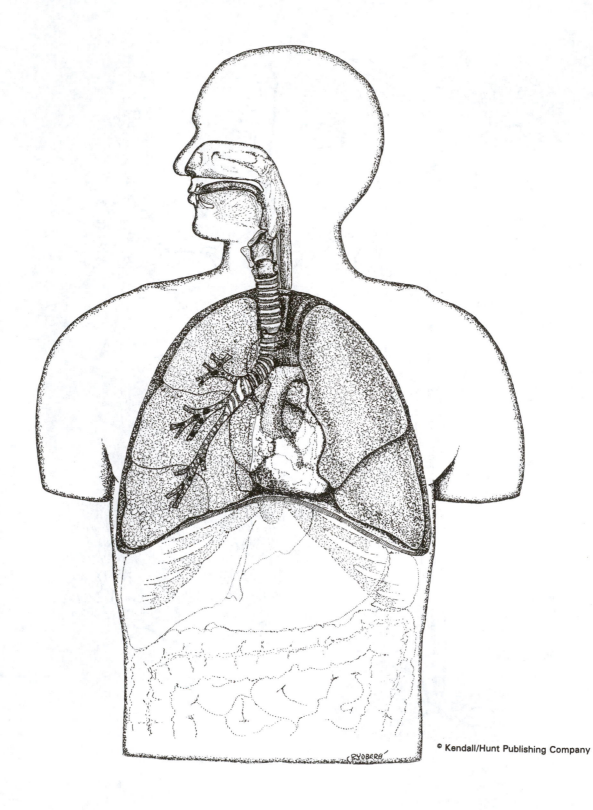

Human respiratory system

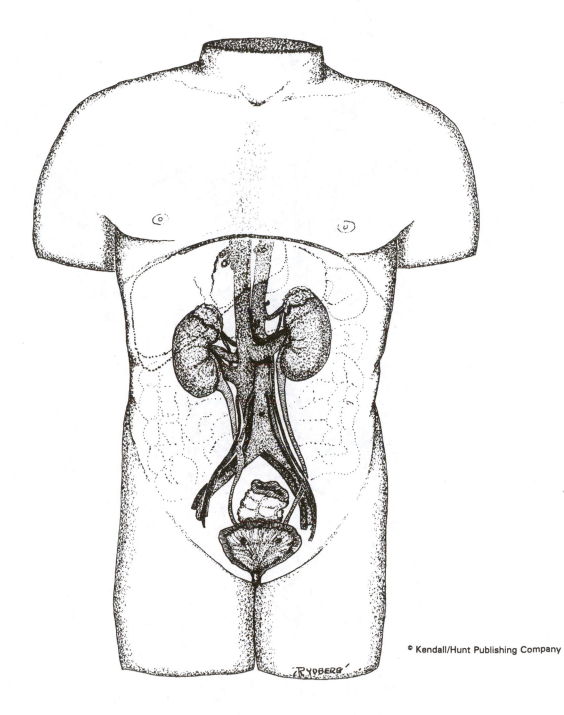

Human urinary system

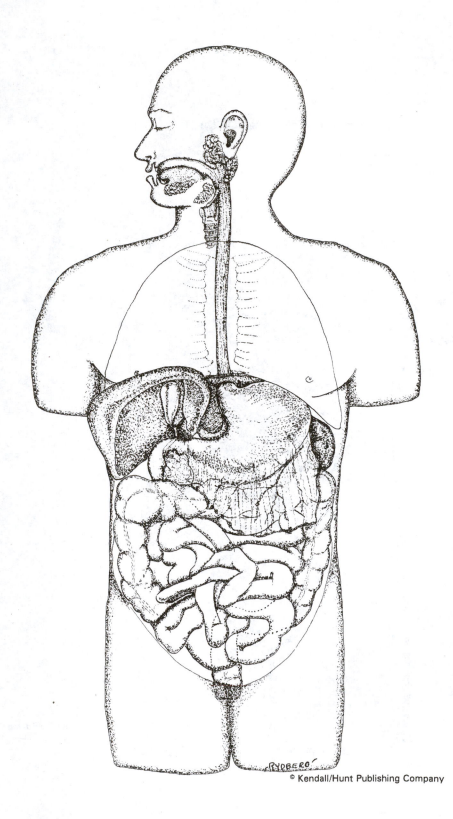

Human digestive system

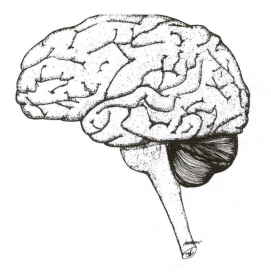

Figure A. Lateral View

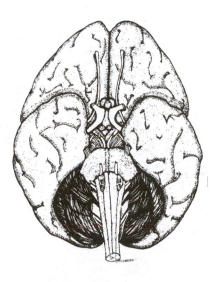

Figure B. Inferior View

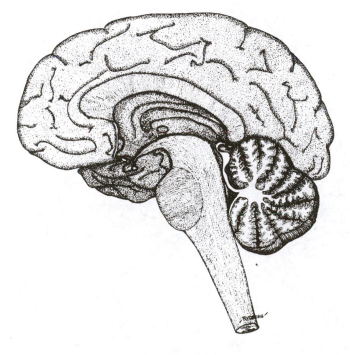

Figure C. Sagittal View

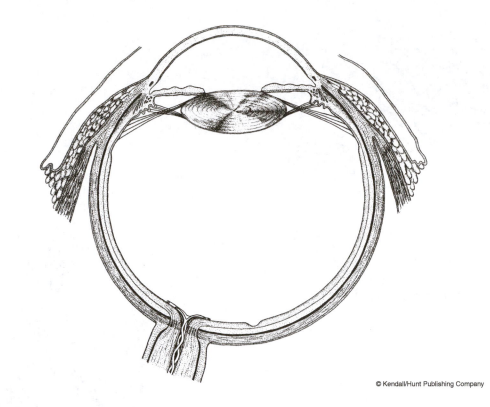

Human eye

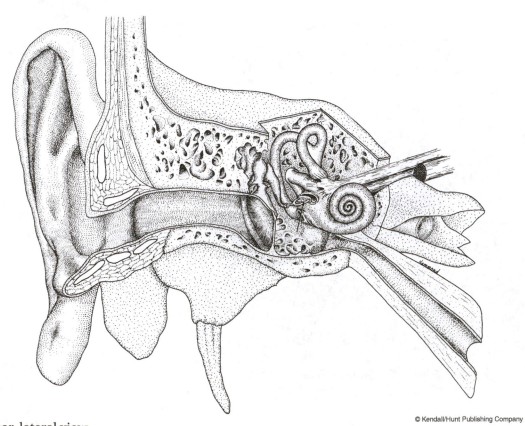

Human ear, lateral view

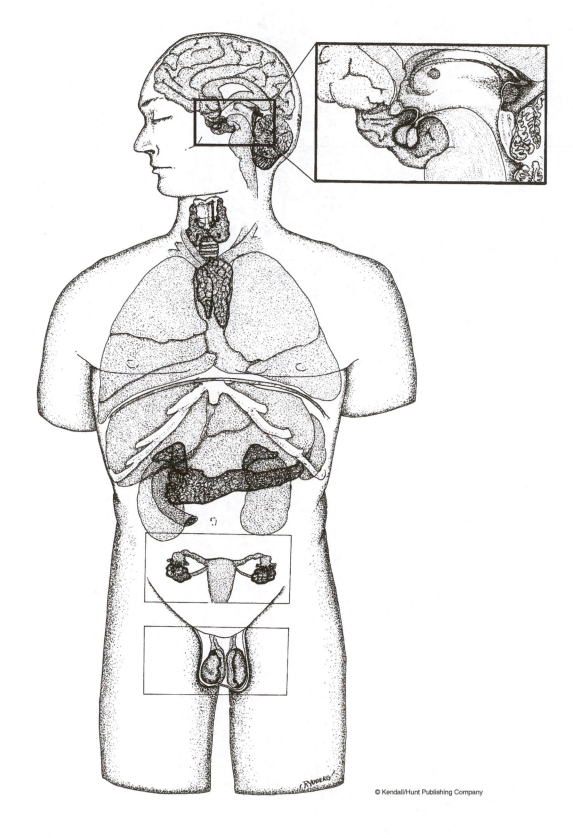

© Kendall/Hunt Publishing Company

Human endocrine system

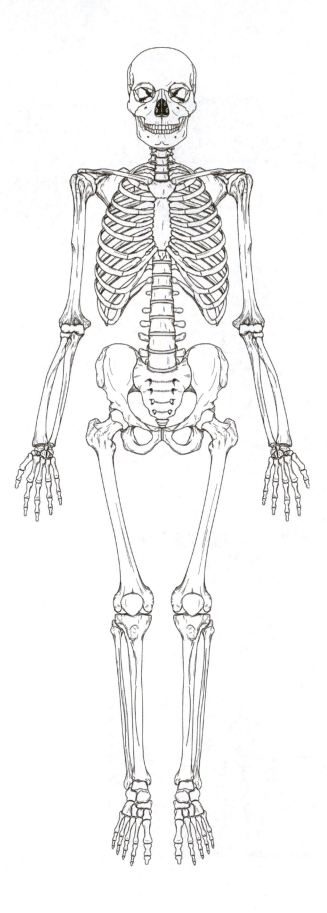

Skeleton, anterior view. © 2003 Mark Nielsen. Art by Jamey Garbett.

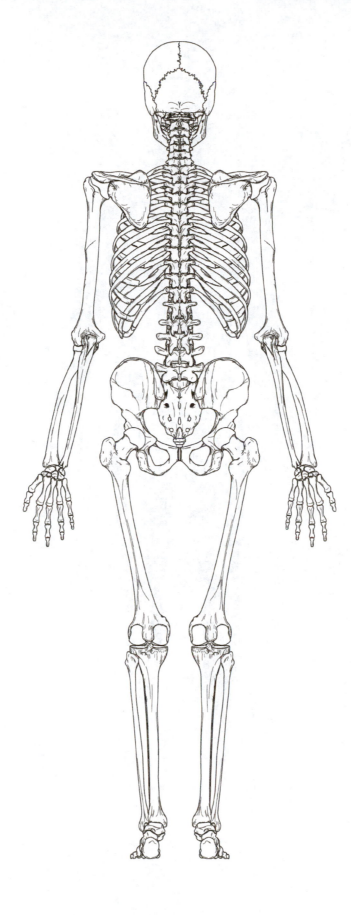

Skeleton, posterior view. © 2003 Mark Nielsen. Art by Jamey Carbett.

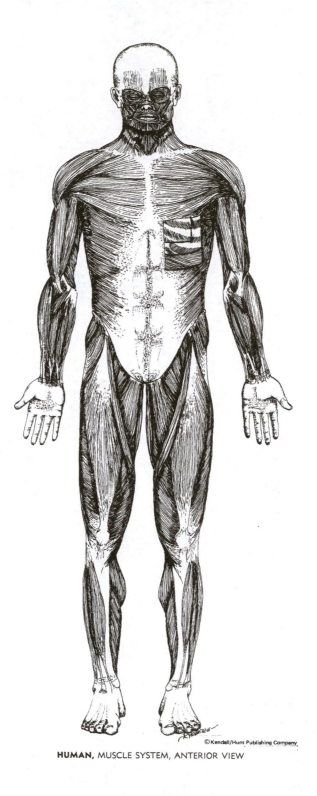

HUMAN, MUSCLE SYSTEM, ANTERIOR VIEW

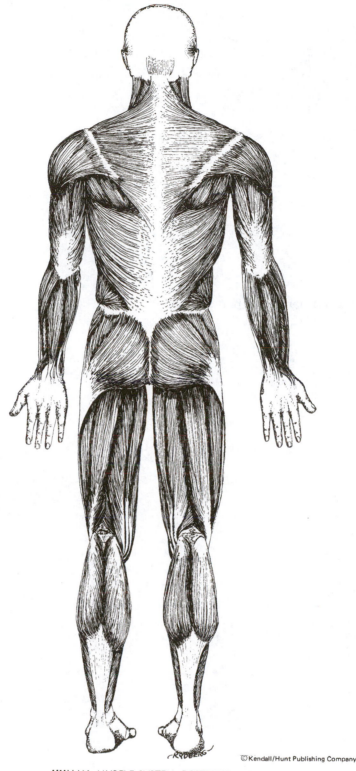

HUMAN, MUSCLE SYSTEM, POSTERIOR VIEW

©Kendall/Hunt Publishing Company

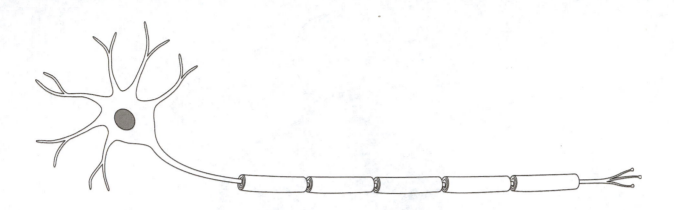